Alexander Vorobyev

A Smoothed Particle Hydrodynamics Method
for the Simulation of Centralized Sloshing Experiments

A Smoothed Particle Hydrodynamics Method for the Simulation of Centralized Sloshing Experiments

by
Alexander Vorobyev

Dissertation, Karlsruher Institut für Technologie (KIT)
Fakultät für Maschinenbau, 2012
Tag der mündlichen Prüfung: 16. Februar 2012
Referenten: Prof. Dr.-Ing. Thomas Schulenberg
              Prof. Dr.-Ing. Hans-Jörg Bauer

Impressum

Karlsruher Institut für Technologie (KIT)
KIT Scientific Publishing
Straße am Forum 2
D-76131 Karlsruhe
www.ksp.kit.edu

KIT – Universität des Landes Baden-Württemberg und
nationales Forschungszentrum in der Helmholtz-Gemeinschaft

KIT Scientific Publishing 2013
Print on Demand

ISBN 978-3-86644-953-4

Acknowledgements

First of all, I would like to express sincere appreciation to my teacher, Dr. Vladimir Kriventsev, for introducing me to the world of CFD and for his guidance throughout my work, starting from our student survey at the faculty of power engineering in Obninsk. His interest in everything new and untraditional has inspired my research on meshless numerical methods and modern computational techniques.

I am very grateful to my supervisor at KIT, Professor Thomas Schulenberg, for his commentary and criticism on my research, which has brought me to a deeper understanding of numerical methods. He has consistently demanded a high-level scientific investigation, and this is also reflected in his careful revision of the manuscript.

I would like to thank all the members of the TRANS group at IKET for the friendly working environment they provide. I especially thank the head of the group, Dr. Werner Maschek, for giving me the opportunity to perform my research at KIT, and for the support he has given me over these years.

I would like to express my special thanks to Dr. Edgar Kiefhaber (KIT), who has helped me greatly in improving the quality of my research. His comments and suggestions have significantly influenced not only the text of this thesis, but also my way of thinking.

I also want to thank Dmytro Trybushnyi (KIT) for the useful discussions on parallel computation, and for his help in solving coding issues.

Last but not least, thanks to Alisa for her support and patience.

Kurzfassung

In dieser Arbeit wird die Smoothed Particle Hydrodynamics (SPH)-Methode als ein numerisches Werkzeug vorgeschlagen und benutzt, um hydrodynamische Prozesse in der Kernenergietechnik zu untersuchen. Diese Methode, bereits weitverbreitet im Bereich der numerischen Strömungssimulation (CFD) und der numerischen Festkörpermechanik, ist im Bereich den thermohydraulischen Simulationen für Kernreaktoren und entsprechenden Sicherheitsanalysen, z.B. Störfall mit einer Kernschmelze, bis jetzt nicht verwendet worden. Die SPH-Methode ist eine netzfreie Lagrange-Partikel-Methode. Wegen der Benutzung von Lagrange-Koordinaten in einem netzfreien Rechenbereich ist sie ein geeignetes numerisches Werkzeug, um CFD-Probleme mit einer komplizierten und sich schnell ändernden Materialverteilung zu lösen. Diese Methode wird als eine zusätzliche Alternative zu den netzbasierten numerischen Methoden betrachtet, die traditionell bei den CFD-Anwendungen in der Kerntechnik verwendet werden.

Die ursprünglich „schwach komprimierbare" SPH-Methode wird verwendet, um einen numerischen Algorithmus zu entwickeln, der fähig ist, die Bewegung der freien Flüssigkeitsoberfläche, vor allem die schwappende Bewegung zu behandeln. Der Algorithmus wird mit mehreren besonderen Viskositätsmodellen und Feststoffgrenzmodellen kombiniert. Statt des künstlichen Viskositätsmodells, das oft in den SPH-Simulationen von freien Flüssigkeitsoberflächen verwendet wird, wird eine Kombination von zwei Viskositätsmodellen verwendet. Dieses kombinierte Viskositätsmodell stellt im Fall der laminaren Strömung eine genauere Beschreibung der Flüssigkeitsviskosität zur Verfügung, und hält dabei für turbulente Strömung die numerische Lösung stabil. Mehrere bekannte numerische Techniken, wie die Dichte-Neuinitialisierungstechnik und die XSPH-Geschwindigkeitskorrektur wurden eingebaut, um die Genauigkeit der ursprünglichen SPH-Methode zu verbessern.

Der numerische Algorithmus wurde in eine neuartige Software umgesetzt, die fähig ist, verschiedene Strömungen mit freier Oberfläche zu simulieren. Um die bekannten Probleme des expliziten Integrationsschemas – hohe Rechenkosten – zu überwinden, wurde die Software mit einem parallelen CUDA Rechenmodul erweitert. Die Parallelisierung des Codes erhöht die Leistungsfähigkeit bis zu 30mal im Vergleich zu einem Standard-PC mit CUDA-fähiger GPU. Der Algorithmus und die Software wurden an Standardtestproblemen validiert und verifiziert.

Die entwickelte Rechensoftware wurde benutzt, um ein aktuelles Sicherheitsproblem schneller Kernreaktoren – die mögliche Rekritikalität wegen der schwappenden Bewegungen der Kernschmelze – zu studieren. Die Geometrie der experimentellen Anlage, die 1992 im KfK (jetzt Karlsruher Institut für Technologie) gebaut wurde, um die Phänomene der schwappenden Flüssigkeit zu studieren, wurde verwendet. Die Ergebnisse der numerischen Simulation geben das in den Experimenten beobachtete Flüssigkeitsverhalten korrekt wieder. Die Genauigkeit der numerischen Lösung, die in dieser Studie mit der SPH-Methode erzielt wurde, ist erheblich größer, als die der bisher mit dem Störfallanalysesystem SIMMER-III/IV erhaltenen Lösung. Die numerischen Simulationen beweisen auch die experimentell beobachtete Tatsache, dass die Bildung der Spitze der Flüssigkeitssäulen sehr empfindlich von der geometrischen Symmetrie abhängt.

Abstract

In this work, the Smoothed Particle Hydrodynamics (SPH) method is proposed as a numerical tool for studying hydrodynamic processes related to nuclear engineering problems. So far this method, widely applied to various problems in computational fluid dynamics (CFD) and computational solid mechanics (CSM), has not been used in the field of nuclear reactor thermohydraulic simulations and the correlated safety analyses dealing with, e.g., molten fuel pools. The SPH method is a meshless particle Lagrangian method. Due to its Lagrangian and meshless nature, it is the perfect numerical tool for solving CFD problems with complicated and fast changing geometry. This method is considered an additional alternative to the mesh-based numerical methods traditionally used in nuclear engineering CFD applications.

The original weakly compressible SPH method is used to develop a numerical algorithm capable of treating free surface liquid motion and, in particular, sloshing liquid motion. The algorithm is accompanied by a number of particular viscosity models and solid boundary models. Instead of the artificial viscosity model, widely used in SPH simulations of free surface flows, a combination of two viscosity models is introduced. The combined viscosity model provides a more accurate description of the liquid viscosity in case of laminar flow, and, at the same time, keeps the numerical solution stable at turbulent flow. A number of known numerical techniques, such as density reinitialization and velocity XSPH correction, are implemented to improve the accuracy of the original SPH method.

The numerical algorithm is implemented as modern software capable of simulating different free surface flows. To overcome the known issues of the explicit integration scheme – in particular, its high computational costs – the software is extended with a parallel CUDA computing module. The parallelization of the code results in as much as a 30-fold performance increase when run on a standard PC with a CUDA-enabled GPU. The algorithm and software are validated and verified against the standard test problems.

Using the developed software, one current problem of fast nuclear reactor safety analyses – that of possible recriticality due to the sloshing motions of the molten reactor core – has been studied. The geometry of the experimental installation, constructed in 1992 at KfK (presently Karlsruhe Institute of Technology) for studying liquid sloshing phenomena, is used. The results of the numerical simulation correctly reproduce the liquid behavior observed in the experimental series. The accuracy of the numerical solution obtained in this study with the SPH method is significantly higher than that obtained with the SIMMER-III/IV reactor safety analysis code. The numerical simulations also confirm the experimental observation that central peak formation is highly sensitive to geometrical symmetry.

Contents

Contents   ix

Nomenclature

B	[Pa]	constant in the equation of state
c_s	[m/s]	speed of sound
C_D	[-]	normalization constant for the smoothing function
d	[m]	initial distance between fluid particles
d_c	[m]	diameter of the liquid column
d_{fb}	[m]	initial distance between fluid and border particles
d_{rod}	[m]	rod diameter
D	[m^2/s^2]	constant in the Lennard-Jones boundary model
D_c	[m]	diameter of container
f	[-]	continuous function
$\vec{F}$	[m/s^2]	external force term
$\vec{g}$	[m/s^2]	gravitational acceleration
h	[m]	smoothing radius
h_c	[m]	height of the liquid column
H_c	[m]	height of the container walls
k	[-]	coefficient defining support domain size
l	[m]	width of the two-dimensional liquid column
m	[kg]	mass
M	[-]	Mach number
$N,\ N_P$	[-]	number of particles
P	[Pa]	pressure
ΔP	[Pa]	excessive pressure
$\vec{r}$	[m]	radius vector
R_c	[m]	distance between rods and container center

$R_{s.domain}$	[m]	radius of the support domain
t	[s]	time
Δt	[s]	time step value
T	[-]	dimensionless time
$\vec{\upsilon}$	[m/s]	velocity
V_f	[m/s]	maximal fluid velocity
V_T	[m/s^2]	viscous term
ΔV	[m^z]	volume associated with a particle
W	[1/m^z]	smoothing function
x	[m]	position of the surge front
X	[-]	dimensionless position of the surge front
Z	[-]	number of dimensions
α	[-]	coefficient in the artificial viscosity model
α_W	[-]	α coefficient for wall-fluid interactions
β	[-]	coefficient in artificial viscosity model
γ	[-]	constant in the equation of state
δ	[-]	Dirac delta function
ε	[-]	coefficient in the velocity smoothing procedure
η	[-]	compressibility factor
λ	[-]	constant in Courant-Friedrichs-Lewy condition
μ	[Pa*s]	dynamic viscosity
ρ	[kg/m^z]	density

Subscript:

| i | value for the given particle |
| j | value for the neighboring particles |

0	reference or initial value
s	smoothed value
ij	difference between values for i^{th} and j^{th} particles
b	value for the border (wall) particle

1 Introduction

The problem of safety in nuclear reactors has been intensively studied from the time of the development of the first reactor designs. In the beginning, the safety problem was not considered a major one – the widespread belief was that a severe accident accompanied by melting of the reactor core would be a very rare event. Over time, several severe accidents occurred at nuclear reactors, but without dangerous consequences for the environment, until the accident at Chernobyl Nuclear Power Plant (NPP) occurred in 1986. This was the most severe accident at a nuclear power unit in history, and together with a previous severe accident at Three Mile Island NPP, demonstrated a strong need for a detailed understanding of the processes that take place in nuclear reactors under emergency conditions, and especially those that occur in the melting reactor core, in order to provide a high level of safety at NPPs. Unfortunately, the year 2011 was marked by a massive severe accident at Fukushima NPP, where four units were seriously damaged by a tsunami wave, leading to the melting of the fuel and radioactive contamination of the surrounding areas. Today it is clear that further successful development of the nuclear energy industry is impossible without deeper knowledge of severe accidents and without the provision of safety guarantees to the public, based on comprehensive analyses of nuclear reactor safety.

1.1 Recriticality Phenomenon in Fast Nuclear Reactors

One of the current problems in severe accident analysis is the problem of corium motion following the melting of the reactor core, which could possibly result in a recriticality event. Such a problem arises under circumstances of a Hypothetical Core Disruptive Accident (HCDA) in the framework of the safety analysis of fast nuclear reactors with liquid metal coolant. Fast reactors have the particular feature that fuel materials in the core are positioned in a configuration which possesses much lower reactivity than the theoretical maximal reactivity for the amount of fissile materials [111]. In fast reactors with sodium cooling, the void sodium effect may also increase the criticality level of the reactor core. For these reasons, the movement of the corium during an accident involving melting of the reactor core may be initiator of a recriticality event with dangerous high power excursions.

Early studies on fast reactor safety were focused on the estimation of reactivity excursion during strong transients, leading to physical disassembly of the core due to buildup of internal core pressure. The duration of the disassembly phase is of order of milliseconds (10 – 100), beginning from reactivity insertion and ending with a transformation of the released heat energy into a pressure build-up, resulting under some circumstances in the destruction of the reactor structures [111].

Later studies showed the existence of an alternative way for severe accidents to develop. Due to the negative neutronic reactivity feedback of the Doppler effect (temperature-induced broadening of neutron cross-section resonances), this accident sequence is characterized by slow core melting. The reactor core melts and forms a large liquid fuel pool confined by frozen fuel, blanket assemblies, and reactor constructions. During corium motion, local fuel compaction may result in a recriticality event and following power excursion, with possible damage to the reactor installation.

The mechanism behind such an accident sequence resulting in recriticality is schematically given in Fig. 1.1. Following [59], in the next paragraph the main stages of the evolution of this type of accident are described.

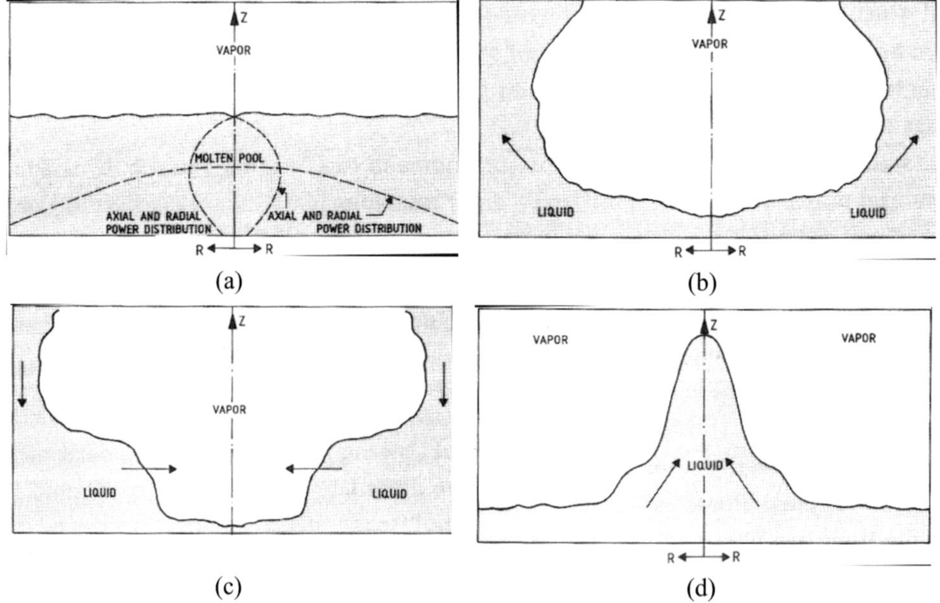

Fig. 1.1. Schematic representation of the sloshing motion of the melted reactor core [59]

As a result of the initiating event, the reactor core undergoes melting into a fuel pool (Fig. 1.1(a)). The local compaction of fissile materials at the center of the pool may trigger a mild reactivity excursion. The consequent energy deposition leads to a pressure build-up in the pool center, which pushes the corium towards the pool periphery (Fig. 1.1(b)). Further, the corium, being driven by gravity, sloshes back towards the core center (Fig. 1.1(c)) and piles up in a neutronically critical or even supercritical configuration (Fig. 1.1(d)).

This "centralized sloshing" motion can lead to a high power excursion; the conditions and processes of this accident sequence have therefore been studied extensively. The main interest of such studies does not lie in the dynamic effects of the sloshing process, such as the maximal value of the impact pressure. It is more important to know the velocity of compaction, the shape of the free surface waves, and the parameters that have an influence on the stability of such complex flow, since it is these conditions that define the reactivity peak parameters during the compaction of the molten fuel.

The sloshing mechanism of the recriticality event is not only specific to severe accidents with core melting in fast nuclear reactors with liquid metal coolant. The sloshing motion can also take place in a pressurized water-cooled reactor, if the core melting results in destruction of the reactor vessel, and following compaction of the corium under the reactor vessel in the reactor cavity [44]. This event sequence is less dangerous, due to the lesser enrichment of the fuel. However, prediction of corium movement is important in designing a suitable core catcher device.

To increase the phenomenological understanding of the sloshing phenomenon, and to provide experimental data for the benchmarks used in the validation of reactor safety analysis codes, a series of experiments has been carried out in KfK (presently Karlsruhe Institute of Technology (KIT)) [59]. The experimental installation consists of a cylindrical container separated by a membrane into two coaxial parts. A sketch of this installation in side view is given in Fig. 1.2. All experimental series were performed for the two general configurations:

> – The inner part is filled with water – the *dam break* problem;
>
> – Both inner and outer parts are filled with water. The water level in the inner part is higher – the *water step* problem.

Experiments were performed with water under normal conditions. The container was opened, so that the environment is air under atmospheric pressure. At the initial moment, the membrane is quickly moved up, resulting in the water column collapsing under the force of gravity. The collapsing water column generates a circular wave that spreads out in the direction of the container walls. Following reflection, a secondary converging wave moves into the container's center, where a central peak is formed. This configuration with a large amount of fissile material compacted in a relatively small volume is most dangerous. Therefore, the main parameter studied in the experiments was the height of the central peak for different geometries.

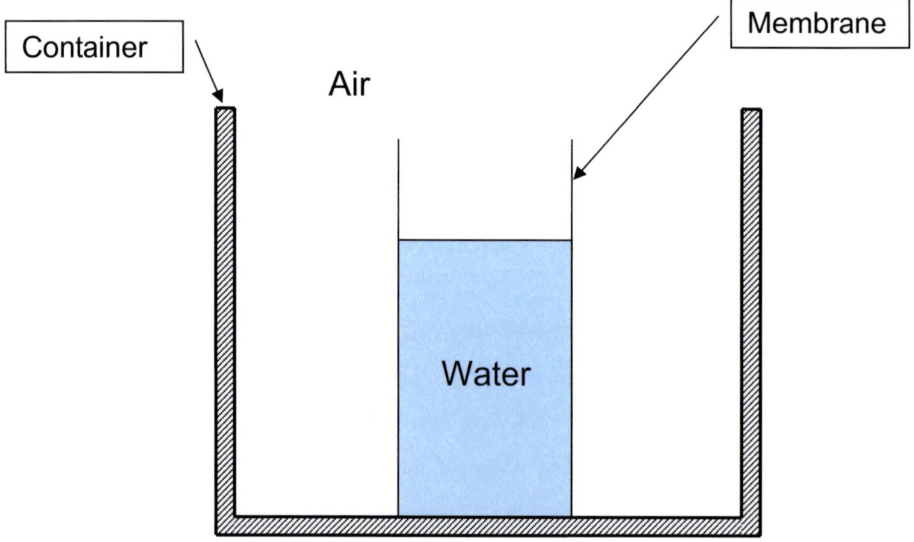

Fig. 1.2. Sketch of the experimental installation used in KfK sloshing motion experiments [59] (side view)

It is well known that converging waves have an inherent instability [93]. Any disturbances leading to violation of the perfectly symmetrical conditions or the coherence of the sloshing process will lead to a reduction in the central peak height. For severe accident analysis, this means a lower power excursion due to a smaller reactivity step. For that reason, the influence of different factors on the stability of the sloshing process was carefully studied in this thesis. Additional elements introduced into the container imitated a situation with a partly melted core, in which some parts of all the fuel assemblies or constructions have not melted.

In Fig. 1.3, the different experimental configurations at the beginning of the experiments are presented:

      (a)   A fully symmetrical configuration with no obstacles in the flow

      (b)   An asymmetrical configuration with no obstacles

      (c)   A symmetrical configuration with a rod bank installed around the liquid column

(d)   A symmetrical configuration with acrylic particles positioned in a ring around the liquid column

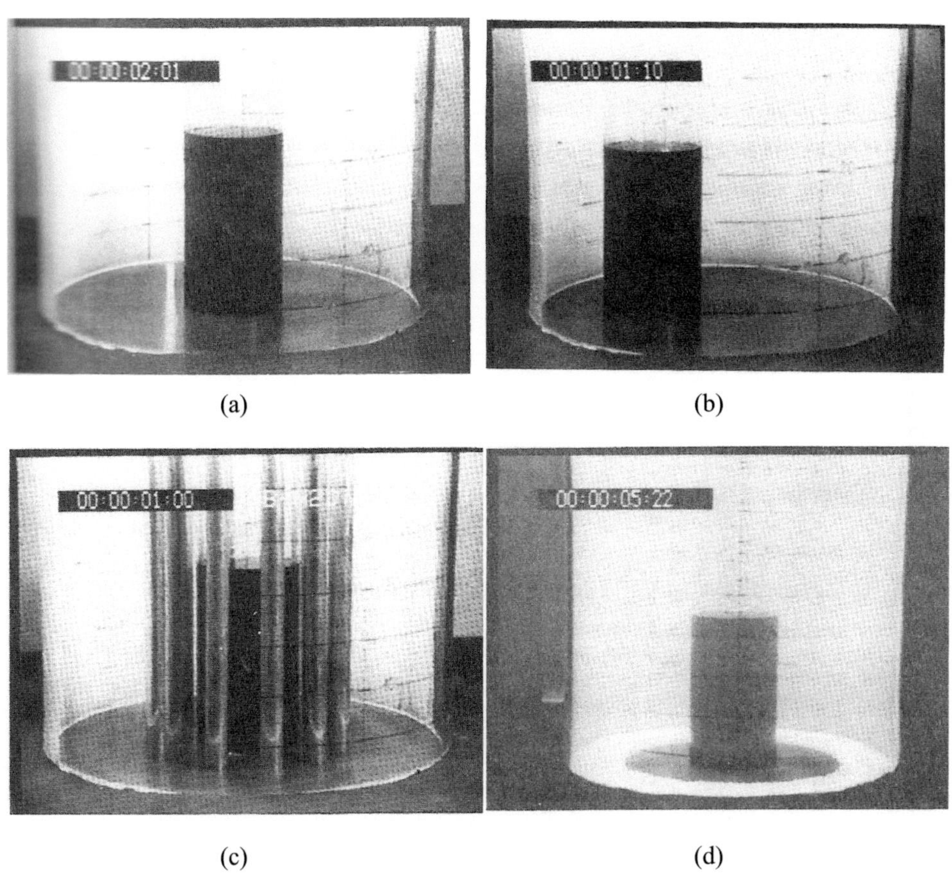

(a)   (b)

(c)   (d)

Fig. 1.3. Experimental installation with different configurations [60]

As has been mentioned, one of the main purposes of these experiments was to provide data for a benchmark exercise for the accident analysis codes. Previous studies on sloshing phenomenon simulation with the SIMMER code [92, 117] have exhibited several limitations and issues related to the Eulerian mesh-based nature of the numerical method. In the following, those issues are discussed in detail.

1.2 Mesh-based Numerical Methods in Nuclear Reactor Simulations

To date, a number of numerical codes for the analysis of nuclear reactor thermohydraulic problems have been developed. Some of these are oriented at the simulation of the normal operation processes of the nuclear reactor. A number of specialized numerical codes for severe accident analysis, including core meltdown, have also been developed [1, 42, 80]. In addition, commercial Computational Fluid Dynamics (CFD) codes, such as ANSYS CFX, FLUENT, and STAR-CD, are applied for numerical simulation of the thermohydraulic processes in nuclear reactors [4, 41, 85].

Traditionally, most of the numerical codes used for reactor safety analysis implement the Finite Differences Method (FDM) and its variations. This method is based on an Eulerian description of the simulated medium, that is, on the use of a static computational mesh. For reactor thermohydraulic problems, it is very common to attach the nodes of the computational meshes to the solid parts of the core, such as fuel assemblies, fuel rods, control rods, cladding, fuel blankets, etc. In this case, the core of the reactor is considered to be a system of parallel channels. The FDM works well when dealing with a "channel-type" geometry in the numerical model. However, in the very unlikely case of a severe accident in the nuclear reactor, the geometry of the computational domain may change significantly as a result of the destructive processes in the reactor core. Moreover, the pool-type design of modern fast nuclear reactor projects poses new challenges to the numerical analysis of their normal operation, and of their behavior under abnormal conditions and in situations of more or less severe accidents.

Summing up, the complex and rapidly changing geometry of the computational domain is a characteristic feature of problems related to the simulation of accidents in nuclear reactors. The existing reactor codes using fixed Eulerian computational meshes have known difficulties in adequately describing such geometries, especially in three dimensions. Unfortunately, the Eulerian approach has several disadvantages when applied to the problems with a rapidly changing geometry of the computational domain [53]:

- It is difficult to treat a complex and irregular computational domain geometry. Usually, costly remeshing procedures are applied when the geometry is changed;
- Using orthogonal grids in FDM results in simplifications of the real existing geometry of the system, because the cell boundaries of an orthogonal mesh cannot perfectly match the boundaries of materials (this feature is mostly encountered in nuclear reactor safety analysis codes, such as SIMMER, while commercial CFD

codes usually employ more flexible computational grids, e.g. fitted curvilinear or unstructured grids);

- The positions of free surfaces, moving material interfaces, and deformable boundaries are difficult to track accurately, since Eulerian methods track the parameters fluxes across cell boundaries;

- The computational mesh is required to overlap all regions in which material can flow, even if the material occupies only a small part of the whole domain at any moment of time. This reduces the resolution of the numerical solution and increases computational costs.

These limitations of the Eulerian approach arise when the FDM-based reactor safety analysis codes are applied to a simulation of the sloshing experiment, which features a fast-moving and complex-shaped free surface of the liquid phase, as well as formation and collapse of drops and jets, etc. The main problem is related to the accurate tracking of the free surface position. (An example of the free surface representation obtained with the SIMMER-IV reactor safety analysis code is given in Fig. 4.5.) It is nonetheless important to note that the latest generations of Eulerian methods are capable of simulating free surface problems, though additional numerical techniques, such as Volume-of-Fluid [37] or Level-Set [94], need to be implemented to track the surface. Although these techniques are quite popular, they come with a number of drawbacks. Application of the additional free surface tracking techniques complicates the computational models, especially for complex multi-component three-dimensional problems, and can lead to undesirable side effects, such as numerical diffusion [34].

1.3 Lagrangian Approach

An alternative way to describe fluid dynamics problems is through the Lagrangian description. Unlike the Eulerian description, where the system of coordinates is spatially fixed, the Lagrangian approach uses moving coordinates attached to the material. As the computational nodes move with the simulated medium, the Lagrangian approach has several attractive features compared with the Eulerian approach [53]:

- Convective transfer of physical parameters such as mass, momentum, velocity, energy, etc., is simulated natively by the movement of the nodes. The convective term is thus excluded from the governing equations;

- The time history of all field parameters can be easily tracked, since the computational nodes are rigidly connected to the moving material

- There is no need for computational mesh generation, which significantly simplifies the handling of problems with complicated geometries of the computational domain;

- Since the computational nodes move together with the simulated material, the free surfaces and interfaces are treated natively, without applying particular tracking techniques;

- In contrast to the Eulerian description, where the computational mesh should overlap all regions that would be occupied by the moving material during the computational period, the nodes in the Lagrangian formulation only represent the volume where the simulated medium is located at the beginning of simulation. This allows using a higher resolution to obtain more detailed information about the simulated system. At the same time, the movement of the computational nodes is not limited by the size of the computational domain, and thus the moving parts of material can be tracked, in principal, as far as desired.

These advantages of the Lagrangian medium description provided a convincing encouragement to the development of numerical methods implementing the Lagrangian formulation of the governing equations. The most popular numerical techniques use moving computational meshes to treat the computational domain in the Lagrangian manner. For instance, one such method is the Finite Elements Method (FEM), which is especially popular for the simulation of Computational Solid Mechanics (CSM) problems, where deformations of the medium are relatively low, compared with CFD problems (some applications of the finite element method to CSM problems can be found in following works: [6], [7], [48]).

However, Lagrangian mesh-based methods suffer from the necessity to follow the medium as it moves, which is often the case in fluid dynamics applications where convection takes place. The tracking of the moving fluid results in highly distorted computational mesh cells, which affect the accuracy of the numerical approximations. The accuracy may be restored by periodical remeshings of the computational domain. For particular problems, such as Taylor-Couette flow where a viscous fluid is confined between two rotating cylinders, the application of the remeshing procedure must be done not only to maintain accuracy, but also to prevent overlapping of the computational cells and the destruction of the numerical solution.

On the other hand, the periodical interpolation of field values into new computational nodes using their values from the old computational nodes also has a negative impact on accuracy [55]. Taking into account the fact that the meshing procedure is an issue for complicated geometries, the limitations of the Lagrangian mesh-based methods in fluid dynamics problems involving a fast moving medium become obvious.

To fully exploit the advantages of the Lagrangian description, it is reasonable to avoid connections between the computational nodes for the approximation of the flow field variables. This can be achieved by applying meshless numerical methods, and, in particular the application of a meshless particle method to the problem. Such methods were originally developed to deal with open boundary fluid dynamics problems, such as astrophysical problems like star formation [57], or plasma modeling problems, where the plasma cloud is considered as a system of particles [77]. (It should be noted that the Particle-In-Cell (PIC) and Cloud-In-Cell (CIC) methods are not truly meshless, as they use an Eulerian mesh together with Lagrangian particles, and hence suffer from the necessity to continually interpolate field variables between particles and nodes of the computational mesh.)

For the simulation of CFD problems in nuclear reactor physics, mesh-based methods have been traditionally applied (see, for example, early work on reactor accident simulations with the SIMMER-I [10], SIMMER-II [92], and RISQUE [5] codes). Recently, the growing interest in the use of meshless methods for nuclear reactor problems has also been noted in a number of works (for example [89, 97, 114]).

Particle methods model the considered physical system as a number of mathematical particles (computational nodes). Each particle can represent a discrete part of the modeled system, like a piece of a solid body, or a part of a continuous medium, such as a volume of liquid. Particle methods need no computational meshes. The interactions between the particles occur following the governing equations, thus define the evolution of the modeled physical system by the movement of particles.

For problems such as sloshing liquid motion or, to use a general term, free surface flows, a capable numerical method should be able to simulate the following physical effects:

    – accurate capture of the free surface and interface movements, including tracking of free surface waves;

    – tracking of pressure waves in the liquid volume;

- formation of liquid drops and jets, their interactions with each other, splashing of liquid medium, and handling problems in which physical substances move out of the computational domain;
- optionally: it is desirable to deal with an easily transformable numerical algorithm, allowing modification of the numerical model to include additional physical effects.

Known numerical methods that match these conditions include the Smoothed Particle Hydrodynamics (SPH) method and the Moving Particle Semi-implicit (MPS) method. The MPS method is an incompressible modification of the SPH method with semi-implicit time integration. In contrast to SPH, the MPS method applies simplified gradient and Laplacian models without taking the gradient of a smoothing function. Another difference is that the MPS method utilizes a smoothing function, which has an infinite value at zero distance [47].

To date, the SPH method has not been used in the field of nuclear reactor thermohydraulic simulations, while the MPS method has been applied to several nuclear reactor applications, namely:

- prediction of the critical Weber number for uranium dioxide drops in water [26];
- simulation of the flashing jets triggered by rapid depressurization during a loss-of-coolant accident in nuclear reactors [27];
- study of the fragmentation mechanisms in vapor explosions [45];
- simulation of drop deposition in annular-mist flow taking place in boiling water reactors [115].

The centralized liquid sloshing experiment has not been investigated using either the SPH or the MPS method. The main idea of this work is to apply the SPH method as one of the meshless particle methods to an actual problem related to nuclear reactor safety analysis – centralized liquid sloshing motion experiment – and to check the advantages and disadvantages of the method. In the next section, a brief introduction of the SPH method is given.

1.4 Smoothed Particle Hydrodynamics (SPH) Method. Short Review

1.4.1 Historical background

The Smoothed Particle Hydrodynamics (SPH) method is a fully mesh-free, Lagrangian particle numerical method. The main ideas underlying the method were independently proposed by Lucy [57] and Gingold & Monaghan [30] in 1977 to solve particular astrophysical problems.

Initially, SPH was described as a statistical method (Lucy even used standard Monte Carlo theory to develop a numerical model [57]), thus estimates of the error were made using Monte Carlo statistics. However, in later investigations (e.g. [31]) it was found that in practice the errors are significantly lower than the Monte Carlo estimates [61]. These early studies demonstrated the high value of the SPH method, and later led to the application of the method for a wide range of different astrophysical problems (e.g. [14], [91]).

The original formulation of the SPH method ([30], [57]) did not provide exact conservation of linear or angular momentum. Further improvements of the formulation were made to ensure conservation of these quantities using the particle equivalent for Lagrange functional [32]. In later publications [62] and [64], Monaghan described the SPH method for simulating compressible flows, paying special attention to the analysis of the method's accuracy and to the evaluation of interpolation errors arising in SPH due to particle disorder. At that time, the SPH method had been applied to compressible confined flow simulations and to various astrophysical open boundary problems. Later, this formulation of the SPH method was applied to free surface liquid flow [72]. This modification, known also as "Weakly Compressible SPH" (WCSPH), is widely used today for simulations of free surface flows, and is also utilized in the present work. (A detailed description of the original WCSPH method, together with the modifications proposed in the context of this thesis, is given in Section 2.)

The formulation proposed and developed by Monaghan was focused on the simulation of inviscid flow, and thus did not approximate the full stress tensor – only the pressure gradient was approximated. To extend the SPH method to material strength problems, Libersky et al. [50] introduced an approximation of the full stress tensor into the "classical" SPH formulation. This was a pilot application of the SPH method to the dynamics of elastic-plastic solids, and its success resulted in a number of applications in related fields: Cleary et al. have investigated several simulations of the high-pressure die casting process [18, 19]; Johnson et al. have studied high-speed collisions of solid bodies and the spreading of elastic waves in solid media [40]; and Benz and Asphaug have simulated asteroid impact events [13].

Simulations of elastic problems have been associated with the discovery of one of drawbacks of the SPH method, namely tensile instability, which is an instability of the numerical solution that occurs when simulating material undergoing tensile stress. This problem rarely occurs in fluid dynamics simulations, but can be very severe in solid body computations. Tensile instability has been investigated by different authors. Swegle et al. [95] concluded that this problem is related to the interpolation procedure used in SPH. Several correction techniques have since been proposed to overcome this issue: Chen et al. [16] have introduced a Corrective Smoothed-Particle Method (CSPM) to address tensile instability and boundary deficiencies. Dyka et al. [28] have proposed a stress-point approach, which is intended to calculate stresses at points shifted from the centers of particles. Monaghan [65] uses artificial stress, which in case of inviscid flow can be represented by an artificial pressure, in order to stabilize numerical solution.

The original SPH method for free surface flows treated free surface problems as single phase systems, though they are actually two-phase flows with a density ratio on the order of 1000. Initially, the SPH method was applied to simulate multi-fluid gravity flows [67]. The results of these studies were very satisfactory, but the density ratios used in such numerical experiments was very low – usually not more than two. Later, the single phase formulation was modified by Colagrossi and Landrini [21] to deal with multi-phase liquid-gas interfacial flows, that is, multi-fluid flows with density ratios of about 1000. Using this formulation, they have successfully predicted the rise of air bubbles inside a water volume, including accurate tracking of the bubble's shape deformation. In addition, they have performed simulations of the water flow released after the breaking of the river dam, taking into account the gas phase, which allowed them to handle effects such as the pressure oscillations inside the air bubble trapped by the wave crest. Further developments of the SPH method in the direction of simulating multi-phase flows have been performed by Hu and Adams in their recent work [38, 39].

The application of the SPH method to realistic interfacial flows has shown the necessity of using the proper surface tension model. The original SPH formalism did not contain any approximation for the surface tension. The idea of the color functions used to assign color values to every fluid particle, and subsequently to calculate curvature of the interface, was introduced by Morris et al. [73] to simulate surface tension in SPH. This model was limited by the confined flows and low density ratios, and thus is not suitable for the simulation of gas-liquid interfacial flows. A new surface tension formulation for multi-phase problems was developed in the work of Adami et al. [2], and this allowed surface tension for the fluid flows with density ratios of up to 1000 to be used.

An extension of the SPH method to turbulent fluid flows has become a logical continuation of its evolution. Violeau et al. [105] have tried to directly approximate the rate-of-strain tensor over the set of Lagrangian particles. Such an approach is not very precise. However, it can give good results in certain particular cases such as the hydraulic jump simulations performed by Lopez et al. [56]. More recently, k-ε model, widely used in mesh-based methods, has been implemented in SPH by Violeau and Issa [106].

At present, the SPH method is undergoing intense development. Different variants of the SPH approximations aimed at increasing the accuracy, stability, and convergence of the numerical solution are proposed by the authors of related studies. (A comprehensive review of SPH development with particular attention to the drawbacks of the method is given by Vignjevic [104]; a recent analysis of the completeness and consistency of SPH has also been carried out by Vaughan et al. [101].) Specialized models for the precise handling of viscosity, heat conductivity, phase change, etc. are still under investigation. The modeling of viscosity is discussed in this thesis, with a brief overview of the most widely used viscosity models given in Section 2.3. Boundary treatment is also an open problem, and this can be an issue for particular types of flow. Some recent suggestions concerning solid boundary simulation can be found in [20] and [70]. In this thesis, the problem of solid wall simulation is discussed in Section 2.4.

The SPH method is currently widely applied to various problems of computational fluid dynamics and computational solid mechanics. Besides the already mentioned works, a number of interesting applications of SPH should be mentioned: simulation of a landslide and the following wave in the real topography of a river valley [83]; analysis of Kelvin-Helmholtz instability in interfacial flows with various density ratios [86]; simulation of material solidification [69], fluid-body interactions [20], heat conduction of liquid-phase diffusion semiconductor crystal growth [17], and underwater explosions, as well as shockwaves generated by the explosions [51, 52]; and many other applications.

The successful application of the SPH method to different fluid mechanics problems, and especially to free surface flows, has defined the choice of the numerical method for the simulation of the centralized liquid sloshing motion, since the last is, in principle, a free surface flow in an axisymmetric geometry. To introduce the mathematical model used in the present work, a brief description of the theoretical basics of the SPH method is given in the following subsections.

1.4.2 Approximation of Functions in the SPH Method

The smoothed particle hydrodynamics method is based on an interpolation technique, which allows the value of any function to be obtained at a given point, using its values at a number of neighboring points. Following Liu G.R. & Liu M.B. [53], the approximation can be described in two steps: the continuous *integral representation* (also referred to as the *kernel approximation*) and the discrete *particle approximation*.

The following integral representation of a function is used:

$$f(\vec{r}) = \int_{\Omega} f(\vec{r}')\delta(\vec{r} - \vec{r}')d\vec{r}' ,$$

(1.1)

where $f(\vec{r})$ is a continuous function, $\vec{r}$ is radius vector, $\delta(\vec{r} - \vec{r}')$ is the Dirac delta function, defined as:

$$\delta(\vec{r} - \vec{r}') = \begin{cases} 1, & \vec{r} = \vec{r}' \\ 0, & \vec{r} \neq \vec{r}' \end{cases} .$$

(1.2)

In the first step, the delta function $\delta(\vec{r} - \vec{r}')$ in equation (1.1) is replaced by the function $W(\vec{r} - \vec{r}', h)$, which is called the smoothing function or smoothing kernel [30, 57]:

$$f(\vec{r}) \cong \langle f(\vec{r}) \rangle = \int_{\Omega} f(\vec{r}')W(\vec{r} - \vec{r}', h)d\vec{r}' ,$$

(1.3)

where h is a smoothing radius, $W(\vec{r} - \vec{r}', h)$ is a smoothing kernel, and the angular brackets mark the approximated value of the function f at the position defined by the radius vector $\vec{r}$.

To ensure the correctness of the approximation (1.3), the smoothing function should satisfy several conditions. The first is the *normalization condition*, which requires the integral of the smoothing kernel over the function domain to be equal to unity:

$$\int_{\Omega} W(\vec{r} - \vec{r}', h)d\vec{r}' = 1 .$$

(1.4)

The second condition is the *delta-function property*, which is observed when the smoothing radius approaches zero [53]:

$$W(\vec{r} - \vec{r}', h) \underset{h \to 0}{\to} \delta(\vec{r} - \vec{r}') .$$

(1.5)

In particular cases, besides the conditions (1.4) and (1.5), additional conditions should be applied to the shape of the smoothing kernel, such as the *even condition* (in which the kernel must be an even function) or the *compact support domain condition* (according to which the kernel must be defined on a compact domain) [53].

In the second step, the continuous integral approximation is converted into the discrete approximation on a number of computational nodes (particles). Following the Lagrangian approach, the continuous medium is represented by a set of particles (computational nodes). Every particle represents a small volume of the medium, and the medium properties at the center of the given volume are assigned to the particle. Further, a continuous field function $f(\vec{r})$ is replaced by a set of numbers f_i – the values of this function at the particle positions. For a discrete number of computational nodes, the integration in Eq.(1.3) can be replaced by a summation, giving the following expression for the function value at the ith computational node (particle):

$$f_i(\vec{r}) = \sum_j f_j W(\vec{r}_i - \vec{r}_j, h)\Delta V_j,$$

(1.6)

where $\Delta V_j = m_j / \rho_j$ is the volume related to the jth computational node (particle).

Substituting the expression for the volume into Eq.(1.6), the final SPH approximation of the arbitrarily continuous function can be derived:

$$f_S(\vec{r}) = \sum_j f_j \frac{m_j}{\rho_j} W(\vec{r}_i - \vec{r}_j, h),$$

(1.7)

where m_j and ρ_j are the jth particle mass and density, respectively; $f_S(\vec{r})$ is the approximated value of the function f at the point defined by the radius vector $\vec{r}$. In general, the summation in Eq.(1.7) is performed over all particles in the computational domain. When smoothing kernels with compact support domains are applied, the summation is limited to a number of neighboring particles.

1.4.3 Approximation of the Function's Gradient

Following Monaghan [64], an approximation of the function gradient $\nabla f(\vec{r})$ is obtained by the use of the gradient of the smoothing kernel. The approximation is as follows:

$$\nabla f_i(\vec{r}) = \sum_j f_j \frac{m_j}{\rho_j} \nabla W(\vec{r}_i - \vec{r}_j, h),$$

(1.8)

where $\nabla W(\vec{r}_i - \vec{r}_j, h)$ is the gradient of the kernel function.

The gradient of the kernel function is calculated using an algebraic derivative of the kernel:

$$\nabla W(\vec{r}_i - \vec{r}_j, h) = \frac{\left(\vec{r}_i - \vec{r}_j\right)}{\left|\vec{r}_i - \vec{r}_j\right|} \frac{\partial W(\vec{r}_i - \vec{r}_j, h)}{\partial(\vec{r}_i - \vec{r}_j)}, \tag{1.9}$$

where $\dfrac{\partial W(\vec{r}_i - \vec{r}_j, h)}{\partial(\vec{r}_i - \vec{r}_j)}$ is an algebraic derivative of the smoothing kernel.

Higher-order derivatives of field functions might be needed for specific problems. For instance, the hydrodynamic simulation of a liquid requires modeling the viscous term in the momentum equation, which depends on the second derivative of velocity. The direct approach for obtaining the second derivative of a function by using the second derivative of the smoothing function can be applied. Unfortunately, such approximations are usually very coarse, and additional models should be implemented to obtain a stable and accurate numerical solution.

1.4.4 Parameters of SPH Approximation

The smoothing radius is a key parameter in the SPH approximation. It defines the distance within which particles interact with each other or, in other words, the distance with a non-zero value of the smoothing kernel (the so called *support domain* of the kernel). In general, a support domain value is a multiple of a smoothing radius value:

$$R_{s.domain} = k \cdot h. \tag{1.10}$$

The value of the constant k is determined by the choice of the smoothing kernel. For a common case with $k = 2$, particles separated at a distance greater than two smoothing radii will have no influence on the parameters at the current point (particle). This is exactly correct when the value of the smoothing function is zero, if the distance to the neighbor point is greater than or equal to $2h$.

The smoothing radius value can be different for every particle, and this can be useful in special cases. A simpler approach gives all particles the same smoothing radius values; this is more computationally efficient, but gives less accuracy for some specific problems.

The choice of the smoothing function has an impact on the accuracy of approximation (1.3). Particularly, Monaghan argued in [64] that – for the following smoothing kernel, based on the cubic spline functions,

$$W(r,h) = C_D \begin{cases} 1 - \dfrac{3}{2}(r/h)^2 + \dfrac{3}{4}(r/h)^3 & 0 < r/h \leq 1 \\[2mm] \dfrac{1}{4}(2 - r/h)^3 & 1 < r/h \leq 2 \\[2mm] 0 & r/h > 2 \end{cases} \quad , \tag{1.11}$$

where C_D is a constant, depending on the number of problem dimensions – the approximation (1.3) has an order of $O(h^2)$. Later in [66], Monaghan demonstrated that higher-order approximations can be obtained with different functions, such as a modified Gaussian:

$$W(r,h) = \frac{1}{h\sqrt{\pi}} \left(\frac{3}{2} - \frac{r^2}{h^2} \right) e^{-r^2/h^2} . \tag{1.12}$$

When the smoothing function (1.12) is applied, the integral approximation (1.3) is accurate to $O(h^4)$. In the same study, it is concluded that the summation interpolant (1.7) can be kept as accurate as desired by means of a corresponding increase of the number of particles involved in the summation. The choice of the smoothing kernel for the present numerical model is discussed in Section 2.5.

In the next section, a detailed description of the numerical model, developed in this study for the simulation of free surface flows, and particularly for centralized liquid sloshing experiments, is given.

2   Numerical Model

2.1 Equations of Fluid Motion in the Formulation of SPH Method

The main equations describing the motion of a viscous incompressible Newtonian fluid are:

- the Navier-Stokes equations (Momentum equation), and
- the Continuity equation.

The Navier-Stokes equations govern the conservation of momentum, while the continuity equation states the conservation of mass. This system of equations can be closed by addition of:

- the Energy equation, and
- the Equation of state.

An important feature of the SPH formulation is the use of the Lagrangian approach to describe fluid motion. According to the Lagrangian description, the above equations are written in a coordinate system rigidly connected to a moving medium. This results in the elimination of the advective term in the momentum equation, so long as the system of coordinates moves together with the simulated medium.

For the free surface flows, such as centralized liquid sloshing motion, the physical processes in fluid flow are quasi-isothermal (obviously an insignificant amount of energy is transformed to heat). This makes it possible to consider the flow as truly isothermic, and thus to exclude the energy equation from the system of governing equations. The choice of the equation of state for the current numerical model is discussed later in Section 2.2.

Using the main approximation principles of the SPH method, the equation of momentum conservation and the continuity equation can be written in the SPH formulation. It is assumed that the computational domain is divided into N small volumes represented by particles. Each particle is assigned mass, density, pressure, velocity, acceleration, and other physical parameters. The number of parameters depends on the particular problem, and can be extended to incorporate more complicated physics in the numerical model.

The equations are written for the ith particle interacting with all other particles representing the system (the j index).

2.1.1 Momentum Equation

The equation of momentum conservation for viscous incompressible fluid takes the following form when written in the Lagrangian formulation:

$$\frac{d\vec{v}}{dt} = -\frac{1}{\rho}\nabla P + V_T + \vec{F} ,$$

(2.1)

where $\vec{v}$ is the flow velocity, ρ is the fluid density, P is pressure, V_T is a viscous term, and $\vec{F}$ is the total volumetric force acting on unit mass. The expression d/dt means the substantial derivative – a derivative of the flow parameters, calculated at a point moving together with a fluid flow.

Using the gradient approximation of the field function (1.8), the SPH approximation for the pressure gradients for the system of particles is as follows [62]:

$$\nabla P_i = \sum_j \frac{m_j}{\rho_j} P_j \, \nabla_i W(\vec{r}_{ij},h),$$

(2.2)

where $P_i = P_i^{abs} - P_0$ is the difference between the absolute pressure at a given point and the initial pressure P_0 , $\vec{r}_{ij} = \vec{r}_i - \vec{r}_j$ is a vector directed from the position of particle j to the position of particle i, h is a smoothing length (smoothing radius), and $\nabla_i W(\vec{r}_{ij},h)$ is the gradient of the smoothing function.

According to Monaghan [62], approximation (2.2) has an important feature. In the case of a spatially constant pressure field, the pressure gradient becomes zero-valued for any configuration of the system of particles. Despite this, Eq.(2.2) does not guarantee conservation of linear or angular momentum. For instance, the non-conservativeness of the approximation can be seen in the problem of a pair of particles with different values of pressure and isolated from any external forces. To ensure momentum conservation, the pressure gradient (2.2) should be "symmetrized," as described in [64].

The mathematical expression for the derivative of the ratio P/ρ can be written in the following form:

$$\frac{\nabla P_i}{\rho_i} = \nabla\left(\frac{P}{\rho}\right)_i + \frac{P_i}{\rho_i^2}\nabla\rho_i .$$

(2.3)

Applying approximation (1.8), we can write the following expressions for the gradients of $\left(\dfrac{P}{\rho}\right)_i$ and ρ_i:

$$\nabla\left(\frac{P}{\rho}\right)_i = \sum_j \frac{m_j}{\rho_j}\frac{P_j}{\rho_j}\nabla_i W(\vec{r}_{ij}, h), \tag{2.4}$$

$$\nabla\rho_i = \sum_j \frac{m_j}{\rho_j}\rho_j\nabla_i W(\vec{r}_{ij}, h) = \sum_j m_j\nabla_i W(\vec{r}_{ij}, h). \tag{2.5}$$

By substituting expressions (2.4) and (2.5) in Eq.(2.3), and by multiplying the left and right sides of Eq. (2.3) by ρ_i, we get the following equation for the pressure gradient:

$$\nabla P_i = \rho_i \sum_j m_j\left(\frac{P_i}{\rho_i^2} + \frac{P_j}{\rho_j^2}\right)\nabla_i W(\vec{r}_{ij}, h). \tag{2.6}$$

Finally, applying the momentum equation (2.1) for the ith particle constituting the computational domain, and substituting the expression for pressure gradient (2.6), the equation for the particle acceleration becomes:

$$\frac{d\vec{v}_i}{dt} = -\sum_j m_j\left(\frac{P_i}{\rho_i^2} + \frac{P_j}{\rho_j^2}\right)\nabla_i W(\vec{r}_{ij}, h) + V_{Ti} + \vec{F}_i. \tag{2.7}$$

The viscous term V_{Ti} needs careful treatment, and is discussed later in the current section. For gravitational flows, such as centralized liquid sloshing motion, the only external force is gravity. Since the term $\vec{F}_i$ represents an external volumetric force per unit mass, it can be substituted by the acceleration due to gravity $\vec{g}$. Where necessary, additional external forces (e.g. wall-on-fluid interaction) can be added to the last term of Eq.(2.7).

2.1.2 Continuity Equation

The continuity equation, which represents the conservation of mass in fluid flow, is usually written in the following form:

$$\frac{d\rho}{dt} = -div\rho\vec{v} = -\nabla \cdot (\rho\vec{v}). \tag{2.8}$$

Applying this equation to the system of N particles, and using approximation (1.8) to substitute the gradient of velocity by the gradient of the smoothing function, the continuity equation is written as follows in the SPH formulation [64]:

$$\frac{d\rho_i}{dt} = -\rho_i \sum_j \frac{m_j}{\rho_j} \vec{v}_j \cdot \nabla_i W(\vec{r}_{ij}, h).$$

(2.9)

To minimize approximation errors, Eq.(2.9) can be transformed in the same fashion as the equation of momentum in the previous subsection, using the mathematical expression for the derivative of the product $\rho \cdot \vec{v}$, rewritten in the following form:

$$\rho_i \nabla \cdot \vec{v}_i = \nabla \cdot (\rho_i \vec{v}_i) - \vec{v}_i \nabla \rho_i.$$

(2.10)

Applying approximation (1.8), the following expressions for the gradients of $\rho_i \vec{v}_i$ and ρ_i can be derived:

$$\nabla \cdot (\rho_i \vec{v}_i) = \sum_j \frac{m_j}{\rho_j} \rho_j \vec{v}_j \cdot \nabla_i W(\vec{r}_{ij}, h),$$

(2.11)

$$\nabla \rho_i = \sum_j \frac{m_j}{\rho_j} \rho_j \cdot \nabla_i W(\vec{r}_{ij}, h).$$

(2.12)

Substituting expression (2.10) in Eq.(2.8), and keeping in mind the expressions for gradients, (2.11) and (2.12), the final expression for the continuity equation in the SPH formulation is:

$$\frac{d\rho_i}{dt} = \sum_j m_j \vec{v}_{ij} \cdot \nabla_i W(\vec{r}_{ij}, h),$$

(2.13)

where $\vec{v}_{ij} = \vec{v}_i - \vec{v}_j$ is the relative velocity of the ith particle with respect to j^{th} particle.

The continuity equation (2.13) is not the only way to define density in the SPH method. The natural way is to apply approximation (1.7) to the density, giving the following equation for the density at the ith particle:

$$\rho_i = \sum_j m_j W(\vec{r}_{ij}, h).$$

(2.14)

Eq.(2.14) establishes the relation between density at a given point and the particle masses at neighboring points. This approach is simple and very straightforward, but has several major drawbacks when applied to free surface flows (for details see [64]). For this reason, continuity equation (2.13) is used for the simulation of centralized liquid sloshing experiments.

2.2 Equation of State

Originally the SPH method was proposed to deal with astrophysical problems, including the modeling of gas cloud interaction, stars and galaxy formation, etc. [30, 57]. All these problems can be treated with the use of some equation of state for the gas medium, e.g. the equation of state for the ideal gas. Later, the method was extended to incompressible fluids [62]. The main idea is that the real liquid, which is almost incompressible, is approximated by a slightly compressible fluid. The maximal possible compression of such hypothetical liquid is artificially limited by the proper choice of constants in the equation of state. The introduction of the artificial compressibility is aimed at obtaining the pressure gradient which defines the particle motion.

Following Monaghan et al. [72], the equation of state proposed by Batchelor [8] to describe the change of pressure with change of density in real water is used:

$$P = B\left(\left(\frac{\rho}{\rho_0}\right)^\gamma - 1\right),$$

(2.15)

where $\gamma = 7$, ρ is current density, ρ_0 is reference density (defined under initial pressure P_0), and B is a constant, which defines how large the change in pressure with the density change is.

According to [49], the constant B in Eq.(2.15) for water is approximately 300 MPa, which corresponds to water's low compressibility. The large value of this constant, when used in simulation, leads to strong pressure fluctuations, and thus to an extremely small time step value. To overcome this issue, the constant B is set to a lower value, resulting in a higher artificial compressibility of the liquid medium. One widely used approach is to limit possible density fluctuations to the 1% level during simulation time. Introducing a compressibility factor η, and given the fact that the compressibility is proportional to the square of the Mach number, the following relations can be written [64]:

$$c_s = \frac{V_f}{M} \cong \frac{V_f}{\sqrt{\eta}},$$

(2.16)

$$B = \frac{\rho_0 c_s^2}{\gamma},$$

(2.17)

where c_s is the speed of sound corresponding to the chosen compressibility, M is the Mach number, $\eta = 0.01$ compressibility factor, and V_f is the maximal velocity of the fluid medium for the given problem.

Currently, the approach of using equation of state (2.15) in SPH simulations of incompressible fluids is referred to as Weakly Compressible Smoothed Particle Hydrodynamics (WCSPH) [9, 53]. There are other possibilities known for fully treating incompressible liquid with SPH method. For example, the authors of [25] used the *projection method* for the simulation of an incompressible fluid. A fully incompressible fluid model was also developed in the MPS method [47]. The group of SPH approaches which model real quasi-incompressible fluid as fully incompressible is typically called Incompressible SPH (ISPH) [38, 87].

The incompressible approach has higher accuracy, but comes at the cost of high complexity in the algorithm and in its computer implementation. Usually, implicit time stepping is implemented, and this leads to additional CPU load. On the other hand, handling of pressure (density) waves in the liquid volume becomes impossible. When the incompressible approach is applied to pressure wave simulation, the use of an additional term to describe the compressibility of the liquid becomes essential (see for example [116]).

There is another drawback to using projection methods in ISPH: it becomes necessary to track the free surface position, which increases computational costs, but more importantly, diminishes the basic feature of the Lagrangian nature of the SPH method – the representation of the free surfaces without application of any particular algorithm.

2.3 Viscosity Modeling

The viscous term in the momentum equation (2.7) depends on the second derivative of the velocity. In principle, the direct approximation of the second derivative in the SPH formulation can be obtained using the second derivative of the smoothing kernel. For instance, such an approximation of the viscous term was used by Takeda in [96]. However, as demonstrated by Gonzales et al. [33], this particular approximation violates the conservation of momentum. In addition, Monaghan in [66] pointed out that the approximation of the second derivative is sensitive to particle disorder, and can lead to instability of the numerical solution. As a result, simplified models of the viscous term (and of other terms related to high-order derivatives) were developed by several authors.

Early SPH simulations were performed for the astrophysical problems with the absence of viscous forces. Later on, in simulating shock waves in gas cloud collision problems, the necessity to introduce a damping term had become obvious (see for example [112]). The artificial viscosity model was introduced to prevent strong shocks and to stabilize the numerical algorithm. Monaghan & Gingold in [68] showed that the use of the artificial viscosity model (mentioned above), when applied to the shock tube problem, led to unsatisfactory results. An improved version of the artificial viscosity model was proposed by Monaghan [63]. For Eq.(2.7), the artificial viscosity term can be written in the following form:

$$V_{Ti}^{Art} = \sum_j m_j \nabla_i W(\vec{r}_{ij}, h) \cdot \begin{cases} \dfrac{-\alpha \bar{c}_{ij} \mu_{ij} + \beta \mu_{ij}^2}{\bar{\rho}_{ij}}; & \vec{\upsilon}_{ij} \cdot \vec{r}_{ij} < 0 \\ 0; & \vec{\upsilon}_{ij} \cdot \vec{r}_{ij} > 0 \end{cases} \tag{2.18}$$

$$\mu_{ij} = \frac{h \vec{\upsilon}_{ij} \cdot \vec{r}_{ij}}{\vec{r}_{ij}^2 + 0.01 h^2}, \tag{2.19}$$

where α and β are constants, $\bar{c}_{ij} = (c_i + c_j)/2$ and $\bar{\rho}_{ij} = (\rho_i + \rho_j)/2$ are the values of the speed of sound and the density averaged between particles i and j. Currently, the artificial viscosity model with some variations is widely used in SPH simulations (for example [21], [36], [102]).

The term involving α produces shear and bulk viscosity, while the term involving β is used to prevent unphysical penetration of particles approaching each other at a high speed. The second term is highly important in the simulation of shockwaves in fluids; it ensures particles are kept far enough apart, and also stabilizes the numerical solution [64].

The values of constants α and β are usually in the range [0.0; 0.1], although the original values suggested by Monaghan [64] were significantly higher, with $\alpha = 1$ and $\beta = 2$. Series of numerical tests (see [107] and also the comments at Fig. 3.3 on page 43) showed that the water volume simulated with such parameters demonstrates significantly higher viscosity than water under normal conditions. For this reason, the values of these coefficients have been considerably reduced for the simulations (see Table 2.1).

The artificial viscous model guarantees stability for the simulations of high velocity flows and for the simulation of free surfaces. The model accurately conserves angular momentum

[64] but, at the same time, creates high viscous forces and therefore cannot predict correct velocity profiles for low velocity flows.

The implementation of the artificial viscosity (Eq.(2.18)) leads to unphysically high shear viscous forces, which depend on the choice of parameters α, β, and h. The tuning of these coefficients is neither clear nor directly derivable from physical conditions. In case of static or slow motion flows, these unphysical viscous effects can be neglected, but for faster flows a more physical viscosity model is required.

A different estimation of the viscous term in the Navier-Stokes equations based on a simplified approximation of the second derivative of velocity was proposed by Morris et al. [73]:

$$V_{Ti}^{Mor} = \sum_j \frac{m_j(\mu_i + \mu_j)}{\rho_i \rho_j} \left(\frac{1}{|\vec{r}_{ij}|} \frac{\partial W_{ij}}{\partial r_i} \right) \vec{v}_{ij}, \tag{2.20}$$

where μ is the dynamic viscosity, and $\dfrac{\partial W_{ij}}{\partial r_i} = \dfrac{\partial W(r_{ij})}{\partial r_i}$ is an algebraic derivative of the smoothing kernel.

This model of viscosity represents a more realistic and physical description of the viscous term (since it depends directly on the dynamic viscosity), and works well for low Reynolds number flows. This physical viscosity model exactly conserves linear momentum, while angular momentum is only approximately conserved [73].

However, the use of the WCSPH artificial compressibility concept can produce numerical oscillations [29], and can ultimately lead to failure to converge, even when using the semi-implicit time integration scheme. The artificial viscosity model damps such oscillations well, while Morris's viscosity model cannot guarantee that numerical fluctuations are suppressed.

Taking advantage of the viscosity models above, a combination of both was proposed in [108]. In order to prevent non-physical high shear viscosity being produced by the artificial viscosity model, the constant α is set to zero. The rest of the right-hand side term in Eq.(2.18) is added to the Morris-type viscous term (Eq.(2.20)) in order to damp numerical oscillations and prevent particle interpenetration. The final expression for the combined viscous term reads as following:

$$V_{Ti} = \begin{cases} V_{Ti}^{Mor} + \sum_j m_j \dfrac{\beta \mu_{ij}^2}{\overline{\rho}_{ij}} \nabla_i W(\vec{r}_{ij}, h); & \vec{\upsilon}_{ij} \cdot \vec{r}_{ij} < 0 \\ V_{Ti}^{Mor}; & \vec{\upsilon}_{ij} \cdot \vec{r}_{ij} > 0 \end{cases} \tag{2.21}$$

When two particles move away from each other ($\vec{\upsilon}_{ij} \cdot \vec{r}_{ij} > 0$), the viscous term in the momentum equation (2.7) becomes identical to the Morris-type viscous term. Otherwise, when particles approach each other ($\vec{\upsilon}_{ij} \cdot \vec{r}_{ij} < 0$), the value of the viscous term is increased, thus stabilizing the numerical solution.

Concluding, it should be noted that the different viscosity models all have advantages and disadvantages when applied to different type of problems. All three viscosity models described in this section have been implemented in the computer code. The main parameters of the different models of viscosity are grouped in Table 2.1.

Table 2.1. Summary for viscosity models implemented in code

Viscosity Model	α	β	μ
Artificial Viscosity Model (by Monaghan [63])	0.01	0.1	Eq. (2.19)
Viscosity Model (by Morris [73])	0	0	Dynamic Viscosity
Combined Viscosity Model (this work)	0	0.1	Both Eq.(2.19) and Dynamic Viscosity

2.4 Treatment of Solid Boundaries

Unlike channel flow, the problems studied in the present work do not require precise simulation of solid walls. The main requirement for the wall model is to provide a non-penetration condition. From this point of view, walls can be represented by the force acting on the fluid particles. Monaghan in [72] suggested that the force acting on the fluid particle with unity mass (which is in fact the acceleration) be calculated from the wall particle as follows:

$$f(r_{bi}) = \frac{D}{r_{bi}} \left(\left(\frac{r_0}{r_{bi}} \right)^{p_1} - \left(\frac{r_0}{r_{bi}} \right)^{p_2} \right), \tag{2.22}$$

where r_{bi} is the distance between a wall particle and the ith fluid particle, p_1 and p_2 are coefficients whose values are usually set to 12 and 6 respectively, r_0 is the initial distance between wall and fluid particles, D is a constant.

The wall force acts along the line connecting centers of two particles in the direction from the wall to the fluid particle. As was mentioned by Liu [53] the value of D should be of the same order as the square of the maximal particle speed in order to ensure the non-penetration condition for fast moving fluid particles. To save computational time, the force is usually truncated at some cut-off distance.

The potential of the force (2.22) is identical to the Lennard-Jones potential, which approximates the interaction between a pair of neutral atoms or molecules. For this reason, the model of solid boundaries which describes the interaction of fluid particle with wall particle by Eq.(2.22) is usually referred to as *Lennard-Jones type* model.

The wall particles are positioned on the solid border with an interparticle distance smaller than the initial distance between fluid particles, thus preventing penetration of the wall border by the fluid particles. The force (2.22) is included in the momentum equation (2.7) as an external force.

In Fig. 2.1, the interactions between Lennard-Jones wall particles and a fluid particle are schematically represented. The wall particles are shown as circles, but in fact they are just centers of the acting force potential. The dashed-dotted circles indicate the maximum radius in which the force acting from a wall particle to a fluid particle is not zero. In the figure, the ith fluid particle only interacts with wall particles 3 and 4. The fluid particle is located equidistantly from both wall particles, and so the interaction is symmetrical and the resulting force acts orthogonally to the wall. However, such an ideal configuration is not always the case. During simulation, fluid particles are chaotically positioned along the solid wall. The resulting force vector is often not normal to the wall. This fact, together with a rapid increase in the absolute force value and a decrease in the interparticle distance, can lead to instability and even to faulting of the numerical solution.

In [66], Monaghan draws attention to another drawback of the Lennard-Jones-type boundary model: a fluid particle moving parallel to a solid wall will experience a variable force that is obviously unphysical. Such variable disturbances can lead to instability of the numerical solution in near-wall regions.

Finally, one specific undesirable feature of the Lennard-Jones-type model emerges, when the model is used to simulate steady-state liquid in a container. In the near-corner regions (especially in the three-dimensional case) a fluid particle interacts with an increased number of wall particles located on the two or three walls connected to the given corner. This results in a higher force directed along the bisecting line of the angle, which develops micro flows in near-corner regions and even leads to instability of the steady-state numerical solution.

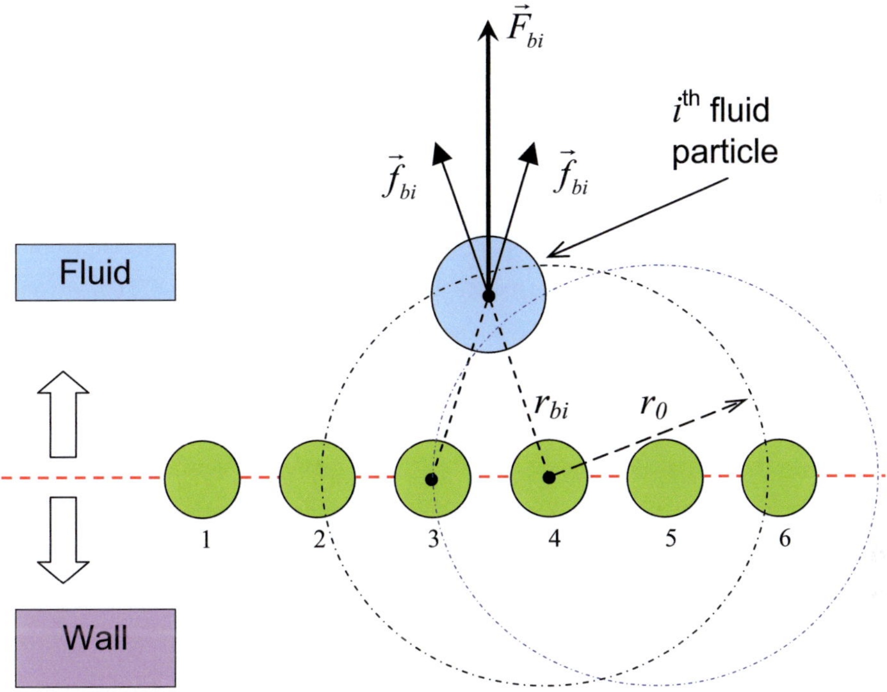

Fig. 2.1. Lennard-Jones boundary particles interacting with a fluid particle

Another method of modeling the solid boundaries has been proposed in the context of this thesis [108]. This method utilizes "fixed" fluid particles, and is rather similar to the "ghost" particle method (see for instance [79]). The idea is to represent the surface of the solid boundary by static fluid particles. The parameters (mass, density, pressure, etc.) of these "fixed" fluid particles are the same as the parameters of the particles representing the liquid medium. During simulation, the interactions of the fluid particles with the wall particles are calculated using Eq.(2.7) and Eq.(2.13). Further, coordinates and other parameters are updated only for the fluid particles, while the parameters of the wall particles remain unchanged.

There are several differences between "fixed" fluid particles and "ghost" particles. Firstly, "ghost" particles are positioned behind the solid boundary, like a reflection of the fluid particles in a mirror. They can also reflect velocities and other parameters of the fluid particles in order to simulate "slip" or "non-slip" boundary conditions. In contrast to the "ghost" particle method, "fixed" fluid particles are placed precisely on the solid surface. Secondly, only one layer of the "fixed" fluid particles is normally used. To ensure non-penetration, they are placed at half the interval of the initial distance of the regular fluid particles. For problems with a relatively low fluid velocity, penetration of the solid wall by fluid particles can be prevented even by increasing the distance between wall particles, thus saving computational time and memory resources.

The "fixed" fluid particle approach, along with the "ghost" particle approach, has a major advantage over the Lennard-Jones type model. The known problem of the SPH approximation is the inconsistency of particle approximation due to the lack of particles near or on the boundary. The "fixed" fluid particles are included in the particle approximation and correct the field parameters. Obviously, the "ghost" particle approach recovers the particle inconsistency at the boundary more accurately than does the "fixed" fluid particle approach. However, series of numerical experiments performed with "fixed" fluid particles applied on the boundary have shown that this coarse model is capable of well ensuring non-penetration, and does not introduce disturbances into the liquid volume.

Using the artificial or combined viscosity model makes it possible to imitate friction between fluid particles and solid walls. For this reason, wall particles are included in the viscous term summation in Eq.(2.7). By tuning the α coefficient, it is possible to check the influence of wall friction on fluid behavior. However, it should be noted that there is no physical background for such modeling of the wall friction.

Summarizing the description of the solid boundary models used in the present algorithm, it is worth noting that different models of solid boundaries each have their advantages and drawbacks in different problems. In Section 3, some specific features of the above-mentioned approaches to simulating solid boundaries are discussed.

2.5 Kernel Function

The quartic spline proposed by Liu et al. [54] is used as a kernel function in the numerical simulations described in Sections 3 and 4. This kernel is defined thus:

$$W(r,h) = C_D \begin{cases} \dfrac{2}{3} - \dfrac{9}{8}\left(\dfrac{r}{h}\right)^2 + \dfrac{19}{24}\left(\dfrac{r}{h}\right)^3 - \dfrac{5}{32}\left(\dfrac{r}{h}\right)^4 & r \leq 2h \\ 0 & r > 2h \end{cases} , \qquad (2.23)$$

where C_D is a normalization constant equal to $\dfrac{1}{h}$, $\dfrac{15}{7\pi h^2}$, or $\dfrac{315}{208\pi h^3}$ for one, two, or three dimensions, respectively.

The kernel function (2.23) exactly satisfies the unity condition (1.4) and the Dirac delta function condition (1.5). It has compact support for itself and for its first derivative. The shapes of the quartic kernel and its first and second derivatives are plotted in Fig. 2.2. For comparison, the cubic spline kernel function (1.11) is also shown.

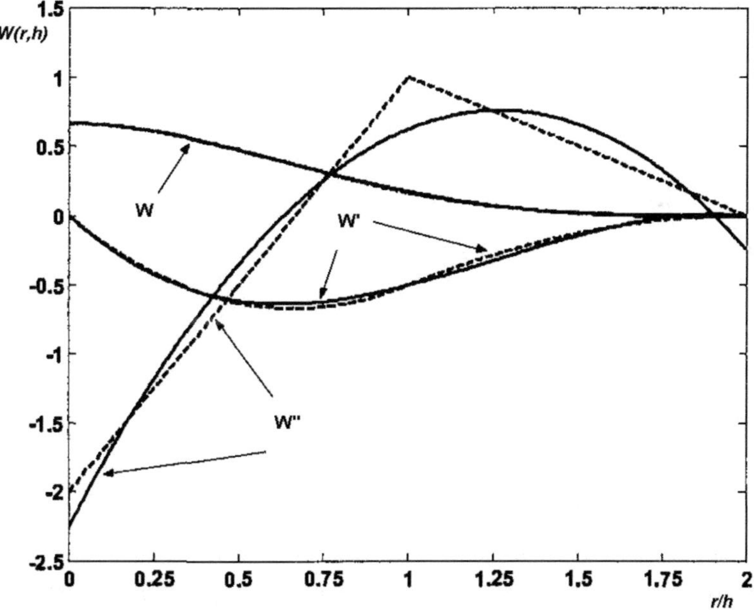

Fig. 2.2. Kernel function with its first and second derivatives. Solid line: Quartic kernel by Liu et al. (Eq.(2.23)); Dashed line: Piecewise cubic kernel (Eq.(1.11)) [53]

The quartic kernel has several advantages over the cubic spline kernel (1.11) widely used in SPH simulations. While the shapes of these two kernels are almost the same, the second derivative of the quartic kernel is smoother than the piecewise linear second derivative of the cubic function. A smoother second derivative generally results in a more stable SPH solution [53]. It is noted in [53] that the quartic kernel yields a more accurate numerical solution than the function defined by Eq.(1.11). Finally, the smoothing function (2.23) consists of one spline only. This allows a simpler and faster numerical algorithm to be developed, especially for parallel calculations on graphical processing units (GPU).

2.6 XSPH Correction

To close the system of governing equations, an equation of motion (written in the so-called XSPH formulation) is added to the momentum equation (2.7), the continuity equation (2.24), and the equation of state (2.15):

$$\frac{d\vec{\upsilon}_i}{dt} = -\sum_j m_j \left(\frac{P_i}{\rho_i^2} + \frac{P_j}{\rho_j^2} \right) \nabla_i W(\vec{r}_{ij}, h) + V_{Ti} + \vec{F}_i, \tag{2.25}$$

$$\frac{d\rho_i}{dt} = \sum_j m_j \vec{\upsilon}_{ij} \cdot \nabla_i W(\vec{r}_{ij}, h), \tag{2.26}$$

$$P = B\left(\left(\frac{\rho}{\rho_0} \right)^\gamma - 1 \right), \tag{2.27}$$

$$\frac{d\vec{r}_i}{dt} = \vec{\upsilon}_i + \varepsilon \sum_j \frac{m_j}{\overline{\rho}_{ij}} \vec{\upsilon}_{ji} W(\vec{r}_{ij}, h), \tag{2.28}$$

where $\overline{\rho}_{ij}$ is the average density of the interacting pair of particles (*i-j*), $\vec{\upsilon}_{ji} = -\vec{\upsilon}_{ij}$ is the velocity of the jth particle in respect to particle *i*, and ε is a constant between 0 and 1.

The equation of motion (2.28) was proposed by Monaghan [63] and is widely used for the numerical simulation of free surface flows. Due to the additional term in Eq. (2.28), the particles are not moved with the velocity obtained from Eq.(2.25), but with a mean velocity weighted with the velocities of the neighboring particles. Using this XSPH correction in the equation of motion has two main objectives:

– to make particle movement more ordered, thus stabilizing the free surface of liquid;

– to prevent the unphysical penetration of particles through each other.

The value of ε is usually chosen from numerical experience. In [63] Monaghan showed that using smoothed velocities does not produce any dissipation. It does however increase dispersion, which means that the velocity field becomes "smoothed" over all neighboring particles. For this reason, the numerical solutions obtained with higher values of ε give less detailed information about simulated flow. And so the smallest value, ensuring stability of the numerical solution, should be chosen.

2.7 Density Reinitialization Technique

The evolution equation for the density (Eq. (2.26)) cannot ensure consistency between mass, density, and area occupied by the particles [21]. To overcome this problem, the density should be periodically reinitialized using a so-called Shepard filter, based on the interpolation technique proposed by Shepard [88]:

$$\rho_i = \sum_j m_j W_{ij}^* (\vec{r}_{ij}, h),$$

$\qquad$ (2.29)

where W_{ij}^* is defined as follows:

$$W_{ij}^* (\vec{r}_{ij}, h) = \frac{W_{ij} (\vec{r}_{ij}, h)}{\sum_j W_{ij} (\vec{r}_{ij}, h) \dfrac{m_j}{\rho_j}}.$$

$\qquad$ (2.30)

The correction (2.29) is formally zero-order accurate. More accurate corrections can also be implemented. For instance in [11], a first-order accurate interpolation scheme based on the moving least squares (MLS) approximation has been proposed. Although the MLS scheme is more accurate, it requires significantly more computational effort. To reduce computation time, the faster correction based on Shepard functions (Eqs.(2.29), (2.30)) is used in this work.

The period of density reinitialization depends on the particular problem. For example, the density reinitialization procedure is performed every twenty time steps for the two-dimensional Dam Break problem, and every thirty time steps for the three-dimensional Central Sloshing simulation.

2.8 Time Integration Scheme

A semi-implicit predictor-corrector time scheme [63] is used for the time integration of the system of the governing equations (2.25)–(2.28).

The time step is controlled by the Courant-Friedrichs-Lewy (CFL) [22] criterion and other stability conditions. Under the assumption of weak compressibility, the "numerical" speed of sound defines the velocity of pressure propagation. Taking into account that the average distance between two neighboring nodes is equal to the smoothing radius h (as is true for the model described in the present study), the CFL condition can be written in the following form:

$$\Delta t_1 \leq \lambda_1 \frac{h}{c_s} = \lambda_1 \frac{h\sqrt{\eta}}{V_f}, \tag{2.31}$$

where c_s is the speed of sound, η is the density fluctuation value, V_f is the maximal fluid velocity during simulation time, and λ_i are the extra limiting coefficients. Usually the CFL condition plays a dominant role in limiting the time step value. Therefore, the choice of the proper value of the speed of sound is of great importance. The speed of sound should be high enough to ensure an acceptable value for density fluctuation. At the same time, a higher speed of sound leads to smaller time steps, and increases the computational cost.

The weakly compressible approach allows the use of even a realistic value of the speed of sound. For a speed of sound of ~1400 m/s and a smoothing radius value of 0.002 m (as used in the sloshing simulation described in Section 4.2), criterion (2.31) yields the time step value of order 10^{-7} s, which is obviously impractical for numerical calculations.

If the artificial viscosity model is applied, the CFL condition (2.31) can be rewritten [64] as:

$$\Delta t_2 \leq \lambda_2 \frac{h}{c_s + 1.2\alpha c_s + 1.2\beta \, \underset{i,j}{Max}(\mu_{ij})}, \tag{2.32}$$

where Δt is defined by the highest value of μ_{ij} (defined by Eq.(2.19)) among all (*i-j*) pairs of the interacting particles.

In the Morris viscosity model, the other time step limit, related to viscous diffusion, should be employed [73]:

$$\Delta t_3 \leq \lambda_3 \frac{h^2 \rho}{\underset{i}{Max}(\mu_i)},$$

(2.33)

where $\underset{i}{Max}(\mu_i)$ is the maximal value of the dynamic viscosity over all particles. The difference in the density values of different particles can be neglected as they are limited to ~1% (see Section 2.2). For the problem studied in Sections 3 and 4, criterion (2.33) predicts higher time-step values than the CFL condition (Eqs.(2.31), (2.32)). It becomes dominant only at high resolution (small h) or large viscosity (big dynamic viscosity μ), neither of which are the case for the abovementioned problem.

In addition, there are limiting conditions associated with the magnitude of the particle accelerations:

$$\Delta t_4 \leq \lambda_4 \sqrt{\frac{h}{\underset{i}{Max}(a_i)}}$$

(2.34)

and the relative velocities between particles:

$$\Delta t_5 \leq \lambda_5 \frac{h}{\underset{i,j}{Max}(\upsilon_{ij})},$$

(2.35)

where $\underset{i}{Max}(a_i)$ is chosen as a maximal acceleration value over all particles, and $\underset{i,j}{Max}(\upsilon_{ij})$ is a maximal value of relative velocity among all (i-j) pairs of interacting particles.

Finally, the time step size is chosen for every time step as a minimal value of the conditions (2.32)–(2.35):

$$\Delta t = Min(\Delta t_2, \Delta t_3, \Delta t_4, \Delta t_5).$$

(2.36)

The constants λ_1 to λ_5 are usually taken to be in the range 0.05–0.4 [63, 64, 73], but the exact values depend on the model parameters. For example, as shown by Morris [73], different kernel functions can have different "effective" resolution lengths for the same smoothing length. In such cases, slightly smaller values for the coefficients λ_1 to λ_3 may be required. For potentially unstable problems, smaller values of the coefficients λ_4 and λ_5 are more preferable.

In the simulations described in Sections 3 and 4, the values of the limiting constants are set to $\lambda_1 = \lambda_2 = \lambda_3 = 0.25$, $\lambda_4 = \lambda_5 = 0.10$

2.9 Neighboring Particles Search Algorithm

An important step in any SPH algorithm is searching for interacting pairs of particles. As noted in Section 1.4.4, interactions between particles occur only within the influence length (1.10) of the smoothing kernel. Usually the influence length is a multiple of the smoothing radius value. For instance, the kernel function (2.23) has an influence length equal to the smoothing radius multiplied by two.

The simplest way to detect neighboring particles is the *all-pair search* algorithm. Generally speaking, this algorithm does not actually search for interacting pairs of particles, but instead checks all pairs of particles for potential interactions. Since the all-pair search approach is carried out for all N particles, and since for every given particle the calculation of interparticle distance is performed N times, it is clear that the complexity of this approach is of order $O(N^2)$ [104]. The search process is repeated at every time step, therefore the use of the all-pair search approach results in huge CPU time cost. It is worth mentioning that, in some particular cases, this algorithm may be preferred. This is the case for the simulation of small systems on graphics processing units (GPU), where conditional switching in the algorithm triggers branch-by-branch execution and thus slows overall code performance. Additional details on the GPU calculations are given in Section 5 and in Appendix B.

A more effective way to find the interacting pairs of particles is to employ the *linked-list search* approach [53], which has become widely used in SPH simulations. The main idea of this approach is to use an additional mesh for searching for neighboring particles. The possibility of reducing the number of computations performed every time step by using an additional mesh was originally noted by Monaghan & Lattanzio in [71]. They suggested using an additional mesh to calculate gravitational forces, and further using the same mesh to determine the nearest particles. In some sense, this was a coupled particle-mesh-based approach, which differs from the fully meshless SPH method.

It is important to distinguish the additional mesh used in the SPH method from the computational meshes used in mesh-based numerical methods. The main differences are as follows:

– The nodes or cells of a computational mesh are used as approximation points, while the cells of the additional mesh are not. They are used only to define the spatial distribution of particles, thus identifying interacting pairs;

– The additional mesh has a simple Cartesian geometry, which does not fit the geometry of the computational domain. In contrast, the computational mesh in mesh-based methods should well fit the computational domain, and so special algorithms of mesh generation are needed.

As long as the additional mesh is used only for searching for the nearest particles, but not for the calculations, the SPH method continues to be truly meshless.

The pair-detecting algorithm with an additional mesh works as follows. First, a simple Cartesian mesh is superimposed over the computational domain. The mesh spacing depends on the influence length (1.10) of the kernel function chosen for the simulations. When the smoothing length is constant for all particles, the cell size is $k*h$. For the case with a variable smoothing length, the cell size may not be smaller than the higher value of the smoothing length over all particles.

The additional mesh, once created, remains unchanged throughout the simulation. For each particle, the corresponding cell is determined at each time step. The resulting list of particles for every cell is stored in memory. Then, for a given particle, only particles located in the same or neighboring cells are checked for possible interaction. The overall number of cells to be checked during the neighboring particle search depends on the number of dimensions. For one, two, and three-dimensional spaces, the number of search cells is respectively 3, 9, and 27. The use of the linked-list algorithm means that the number of required computations per time step is proportional to the total number of particles used for the simulation [71].

In Fig. 2.3, the linked-list search algorithm for a two-dimensional problem is illustrated. The solid-lined square bounds the cells where particles are checked for possible interactions with the given particle (red). Only particles located within the influence domain are involved in interactions (interacting pairs are marked with arrows).

Although the linked-list algorithm is highly efficient, especially for problems with a large number of particles, it cannot handle effectively problems with variable smoothing lengths. For such problems, the *tree search* algorithm, which has a complexity of order $O(NlogN)$, can be used [53].

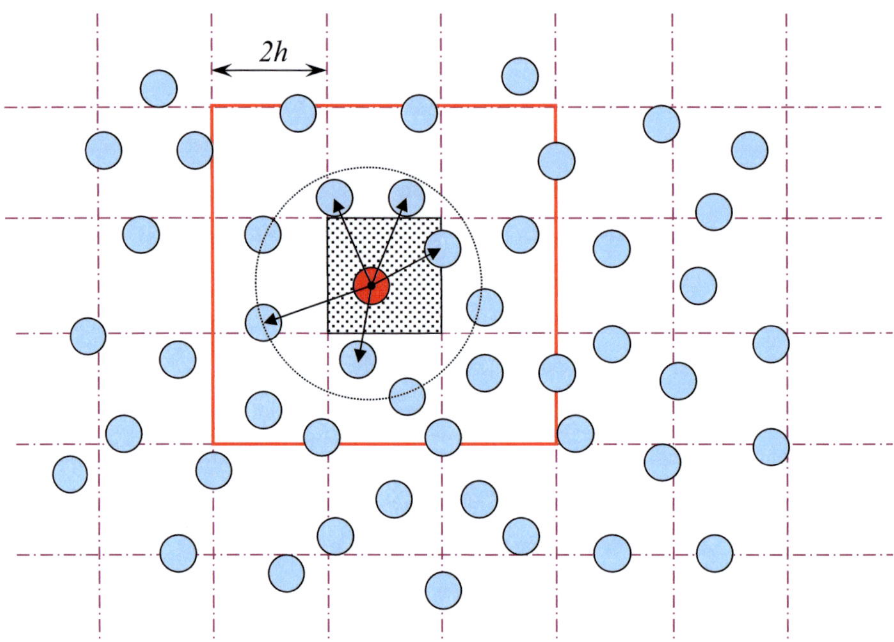

Fig. 2.3. Additional mesh used in linked-list algorithm for searching for interacting pairs of particles (two-dimensional space)

2.10 Concluding Remarks

In this chapter, the developed numerical model based on the SPH method has been described in detail. The model involves a number of known specialized techniques which improve the stability and accuracy of the original SPH approximation. A combined approximation for the viscous term in the momentum equation has been introduced to allow approximation of the fluid viscosity by using a dynamic viscosity, which maintaining the stability of the numerical solution at the same time.

The numerical model has been implemented in specialized computer software – SimSPH [90]. Details of the computer implementation of the SimSPH application can be found in Appendix A.

3 Model Validation

3.1 Introduction

The validation assessment is an important part of the development of any numerical code. The American Institute of Aeronautics and Astronautics (AIAA) defines validation as:

The process of determining the degree to which a model is an accurate representation of the real world from the perspective of the intended uses of the model. [3]

Clearly no mathematical model is a complete description of the real world. Some simplifications usually lie at the base of the model, binding its application to specific problems under certain conditions. The numerical code described in Section 2 has been developed for a particular hydrodynamic problem: centralized sloshing experiments, with a further extension to general free surface gravitational flows.

The postulated goals of this study define the choice of the test problems used for the validation:

- a static container filled with liquid under the force of gravity;
- the dynamic problem of a breaking dam and the subsequent liquid motion (the "Dam Break" problem);

The validation strategy is mainly based on a comparison of the numerical code predictions with experimental observations, while the verification process includes consistency and balance checks, such as checks for mass conservation, hydrostatic pressure distribution, etc. [103]. So far as such checks have been performed, it would be correct to say that this section is related to the "verification & validation" process.

The numerical simulations of the test problems described in this section are not only aimed at verifying and validating the numerical model, but also at helping to examine the behavior of different models, such as the solid wall model and the viscosity model.

3.2 Static Problem. Simulation of Hydrostatic Pressure Field

The distribution of the pressure in a stationary liquid column is quite trivial to simulate with traditional mesh-based approaches. The meshless methods, and especially the particle methods, have several features which should be taken into account when simulating static problems. Unlike mesh-based approaches, the computational nodes (particles) have no connections and thus are free to move. When the weakly compressible approach (see Section 2.2) is applied, the density (and respectively the pressure) can vary with time. As a result, particles oscillate following waves of pressure. This feature complicates the calculation of the hydrostatic pressure. In case of a container filled with liquid, it is important to use a stabilizing term to prevent the disturbances of the liquid volume due to the boundary forces. Without damping, both unstable numerical solutions and quasi-stable fluctuating states of the particle system can be produced. In the quasi-stable state, particles can move and oscillate, leading to deviations of the local parameters, while the integral (smoothed over the neighboring particles) parameters, like the hydrostatic pressure profile, remain nearly constant.

The proper balance between disturbances from the wall particles and damping by the viscous term should be found to accurately simulate the gravitational flows. This problem has been studied in one and two-dimensional formulations, with the application of different solid walls and viscosity models.

3.2.1 One-Dimensional Hydrostatic Pressure Test

For the one dimensional case, the computational domain (one meter in height) is represented by a single vertical row of twenty particles. The Lennard-Jones type boundary particle (see Section 2.4 for the details) and the "fixed" fluid particle are used as a one-dimensional "wall." All fluid particle properties are assigned constant values corresponding to water under normal conditions. The particles are positioned equidistantly, with the interparticle distance equal to the smoothing radius h. The masses of the fluid particles are defined as $h*\rho$. The interactions between particles are calculated only for the vertical components of velocity, acceleration, etc. In Fig. 3.1, the computational domain at the initial state is schematically presented. At the beginning of simulation, all particles are assigned zero velocities. Under the influence of gravity, the fluid particles begin to move and interact with the fixed border particle and with each other. Due to the presence of viscous diffusion, the system of particles loses energy and goes to steady-state through the oscillation mode. When Morris's viscosity model is applied with the Lennard-Jones model for the wall representation, the system becomes unstable. The top fluid particle experiences incremental oscillations of pressure, and is pulled away from the particle column, thus breaking the continuity of the liquid volume. Using the "fixed" fluid particle to represent a solid wall, all the viscosity models described in Section 2.3 can stabilize the numerical solution and suppress particle oscillations.

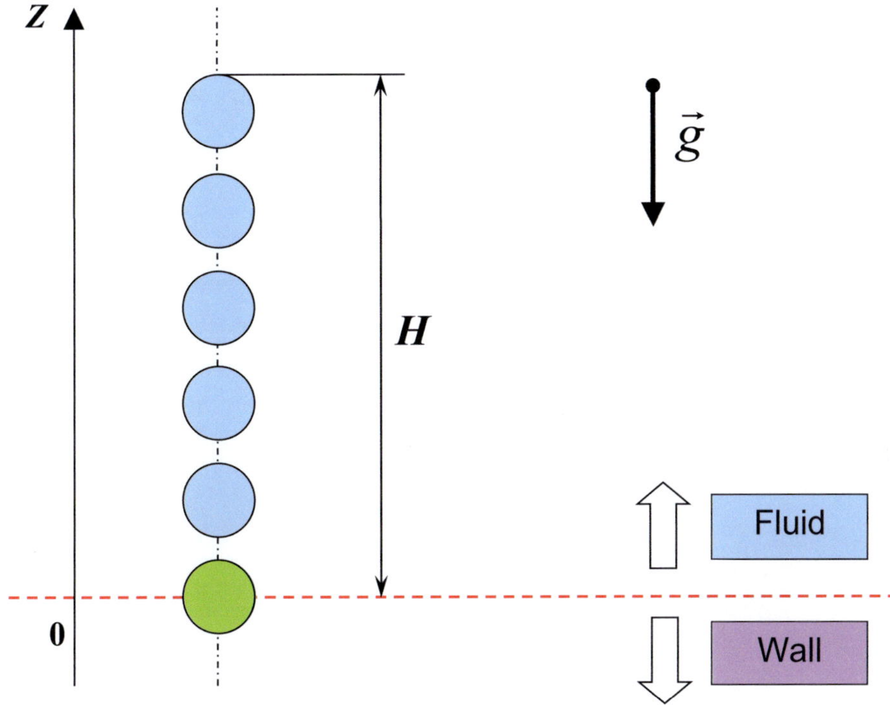

Fig. 3.1. One-dimensional static test case (sketch)

The distances between particles are changed according to the pressure increase. The final compression of the fluid volume is about 1% for both the Lennard-Jones and the "fixed fluid" types of boundary particle. When using the Lennard-Jones particle for the wall simulation, the fluid column additionally moves down, because the shape of the interaction potential (2.22) leads to a zero force for the initial configuration. The steady state in this case is reached only when the lowest fluid particle is moved to the influence domain of the boundary particle. The offset of the fluid column depends on the choice of value for the constant D in Eq.(2.22).

The resulting steady-state pressure profile after 1 ms of simulation is plotted in Fig. 3.4, together with the pressure profile calculated for the two-dimensional case. The results of both simulations are discussed in the following subsection.

3.2.2 Static Tank Filled with Water (Two-dimensional Problem)

In two dimensions, the computational domain consists of a top-opening rectangular container partly filled with a liquid medium. The container width is 1.5 m and the liquid level is 1.0 m. The side walls of the container are 1.2 m in height, although the exact value is not important for this test problem. The container is an open vessel; the liquid medium is thus under atmospheric pressure.

The liquid medium is represented by a set of fluid particles, initially equidistantly positioned on Cartesian mesh nodes. All fluid particles are assigned constant values of pressure and density, which correspond to atmospheric pressure. The mass of the particle is defined as the product of the initial density and the volume of liquid represented by a fluid particle. The initial configuration of the particle system is shown in Fig. 3.2.

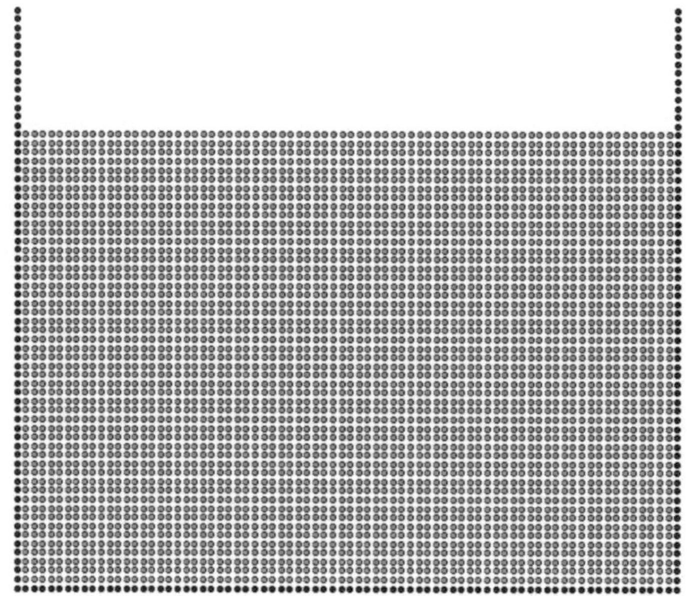

Fig. 3.2. Initial positions of particles (static test problem in two dimensions)

Similar to the one-dimensional case, the particles move down under gravity during the initial phase. As the wall-fluid interaction forces increase, the fluid particles slow down and form a new quasi-stable configuration.

In order to damp the particle oscillations during transients, different approximations of the viscous term in the momentum equation have been used. Depending on the viscosity model in use, the fluid particles either oscillate at their new positions or continue to move with slow velocities. A number of features in the application of each viscosity model are discussed below.

The artificial model of viscosity (2.18) damps the numerical oscillations of the particle system well, and stabilizes the free surface of the liquid volume. Due to the introduction of high viscous dissipation, the particles form specific layered structures in the steady state (see Fig. 3.3 (a)). These structures do not affect the accuracy of the averaged parameters, but they do distort the local values of field parameters, such as pressure, density, velocity, etc.

The Morris approximation of the viscous term (2.20) does not introduce any additional shear viscous forces. As in the case with the artificial viscosity model, the fluid particles occupy new quasi-stable positions and continue to oscillate there after the transient period. The Morris viscosity model damps such numerical oscillations, but not sufficiently. As a result, the particles positioned on the free surface can be accelerated quickly enough to violate the continuity of the liquid medium, thus causing the numerical solution to fail.

The last viscosity model that has been applied to the static problem is the combined model (2.21) proposed in this work. As mentioned in Section 2.3, this model does not introduce excessive shear viscous forces or make the particle distribution more chaotic (see Fig. 3.3 (b)), thus reducing the deviations of the local parameters from the averaged values. This moreover ensures a highly stable numerical solution, as in the case of the artificial viscosity.

It should be noted that the application of the combined model to such a static problem leads to the same results as the application of artificial viscosity with a low or zero value of the α coefficient, since the viscous forces can be neglected. In this case, the model of viscosity plays only a stabilizing role.

It was found that the use of the Lennard-Jones boundary particles in the two-dimensional case leads to disturbances in the near-bottom corner regions. Under certain circumstances, these disturbances result in some fluid particles undergoing extremely high acceleration. These highly accelerated particles can destroy the rest of the system even for low values of the coefficients λ_4 and λ_5 in Eqs. (2.34), (2.35) (and for respectively smaller values of the time step). A more stable numerical solution can be obtained when "fixed" fluid particles are used to represent solid walls.

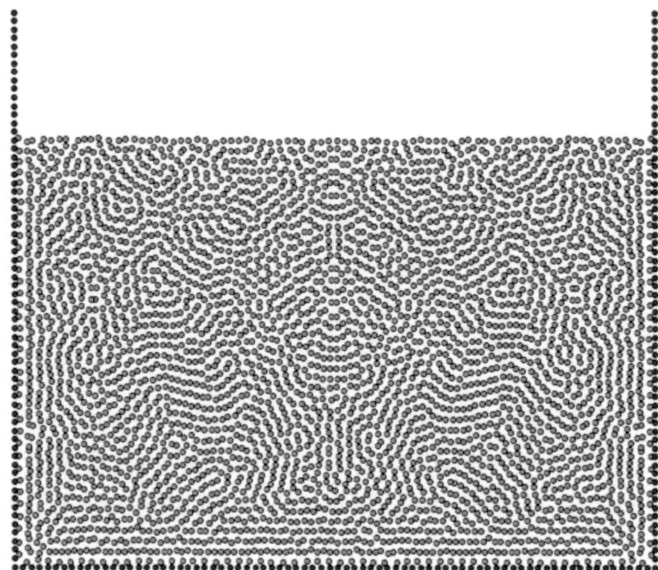

(a) The artificial viscosity model (Eq.(2.18)) leads to the formation of layered structures

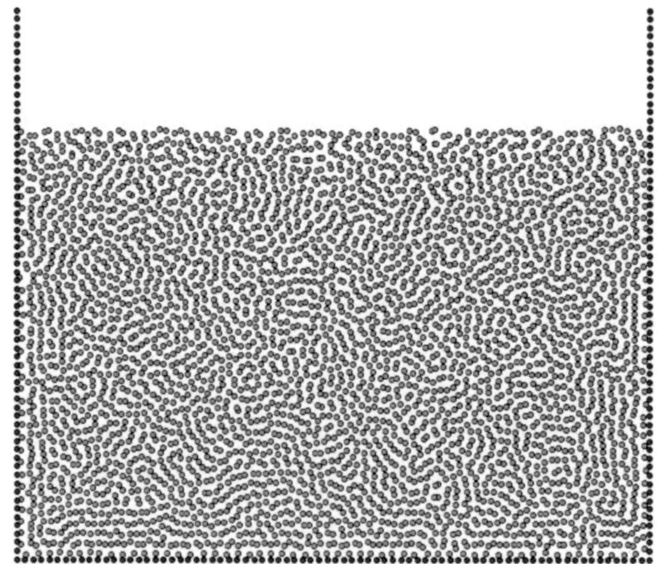

(b) The combined viscosity model (Eq.(2.21)) provides a more random distribution of particles

Fig. 3.3. Two-dimensional static water-filled tank. Steady-state (t = 600 ms)

In Fig. 3.4, the pressure profiles for the one-dimensional and two-dimensional problems are plotted. For the one-dimensional case, the pressure values at all particles are used. In plotting the two-dimensional case, only the pressure values of particles located within the 5-cm innermost layer near the vertical axis of symmetry of the container are considered.

The calculated pressure profile in the one-dimensional case is "saw-shaped". However, the mean pressure line agrees well with the analytical solution for the hydrostatic pressure. The "saw-shaped" pressure profile is a result of using the "fixed" fluid particles with constant parameters. The pressure at the wall does not change during the transient. At the time shown on the one-dimensional case plot, the pressure at the wall particle is still atmospheric, while the pressure at the closest fluid particle is ΔP higher, where ΔP is a hydrostatic component of the total pressure. This pressure drop at the wall region disturbs the whole one-dimensional particle row, leading to the observed "saw-shaped" pressure profile. The deviations of the pressure values from the mean pressure decrease as the free surface is approached.

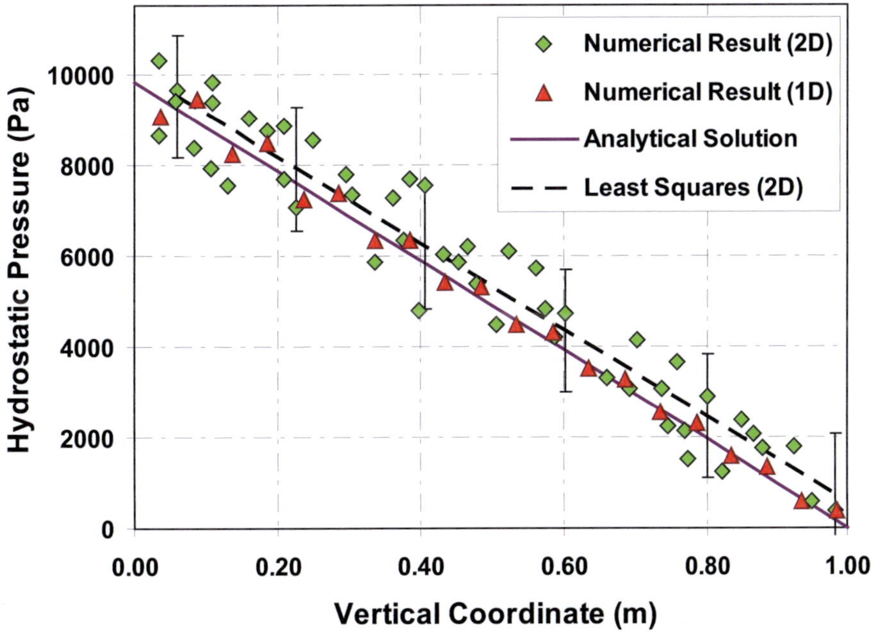

Fig. 3.4. Pressure distribution calculated in one and two-dimensional simulations of a static water-filled tank. The one-dimensional plot shows the pressure 1.0 s after the simulation's start. The two-dimensional plot shows the pressure field 0.6 s after the simulation's start

In the two-dimensional domain, fluid particles can move in both vertical and horizontal directions, resulting in continuous interparticle interactions in the near-wall regions. Due to the momentum exchange, the unphysical "saw-shaped" pressure field is smoothed and dispersed, as shown in Fig. 3.4. The mean of hydrostatic pressure, obtained using the method of least squares, is slightly higher than in the analytical solution. However, the mean agrees with the analytical solution in the range of the standard error. The observed overestimation is due to the additional compression in the near-bottom wall region. This effect becomes less noticeable at higher resolution (higher number of particles). In addition, the application of the density reinitialization technique (see Section 2.7) restores the density field according to the actual particle distribution, and thus reduces the observed deviation. Moreover, for long simulations of static liquid volumes, the use of the density reinitialization technique becomes necessary to avoid the accumulation of numerical errors in the continuity equation (2.26). In Fig. 3.5, the pressure fields for the simulation performed without reinitializing density, and for the simulation with density corrected every 20 time steps, are plotted.

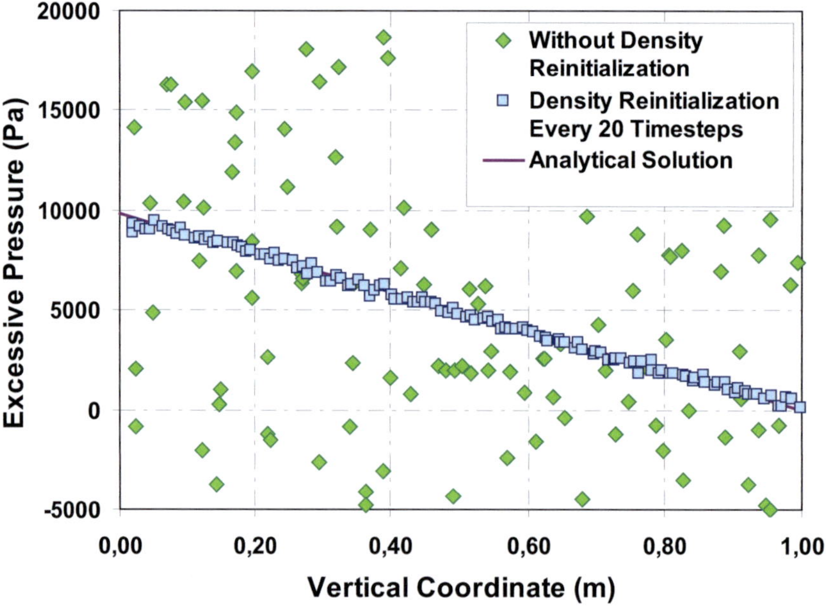

Fig. 3.5. Pressure distribution calculated in two-dimensional simulations of a static water-filled tank with and without reinitialization of density. Both plots show the pressure field 2.0 s after the simulation start

As can be seen from the pressure plots, the application of the density reinitialization technique helps to ensure consistency of the numerical solution, thus providing the correct pressure profile, while the solution computed without density correction completely fails to predict the hydrostatic pressure profile. Longer computations (up to 5 seconds of real time, not shown here) demonstrate a tendency to diverge for the solution without the density reinitialization. At the same time, the solution obtained with periodic correction of density converges to the analytical solution.

3.3 Two-Dimensional Dam Break Test Case

The problem of the collapsing water column caused by a dam failure has been widely studied experimentally, theoretically, and numerically by many researchers (for instance [15, 47, 113]). The primary reason for this interest is that a dam break, if it were to occur, could have great impact on the population and the environment. Irrespective of real dam breaks, an abstract "dam break" problem is a good validation test for numerical algorithms dealing with gravitational free surface flows. The sloshing experiment studied in the present work can also be considered a circular dam break. Since the two-dimensional dam break problem has become a traditional validation test for the SPH algorithms, it seems reasonable to perform the validation of SimSPH code by simulating the dam break test case.

One of the best known experimental studies of the liquid motion subsequent to the breaking of a dam was carried out by Martin & Moyce [58]. A sketch of the experimental installation used in their study is presented in Fig. 3.6. The installation consists of a channel with a rectangular cross-section divided by a thin membrane in two parts. One part is filled with water, while the other is kept dry. The water column has width l and height $h = n^2 l$. The absolute sizes of the water column are not important, as the results can be expressed in dimensionless form. Experimental tests were carried out for different values of n^2.

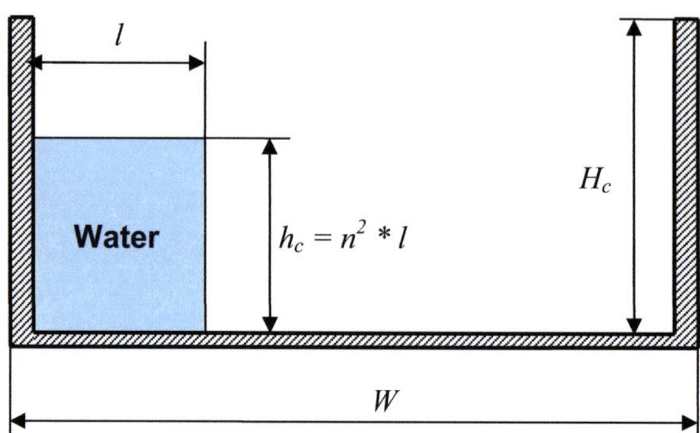

Fig. 3.6. Two-dimensional dam break. Problem sketch (side view)

At the beginning of the experiment, a strong electric current is passed through the paper diaphragm, leading to its release from the fixed position. The motion of the liquid is captured using a fast video camera through the transparent side walls of the channel.

The overall motion of the liquid after release is shown in Fig. 3.7 (a). The main parameters characterizing the liquid motion are the velocity of the surge front and the shape of the free surface. Since the experimental installation used by Martin & Moyce was open at the opposite end, effects such as the reflection of waves from a walls or reverse wave motion were not studied. To obtain data on these phenomena, Koshizuka et al. [46, 47] performed a series of similar dam breaking experiments with a channel closed on both ends. The corresponding experimental snapshots are shown in Fig. 3.7 (b).

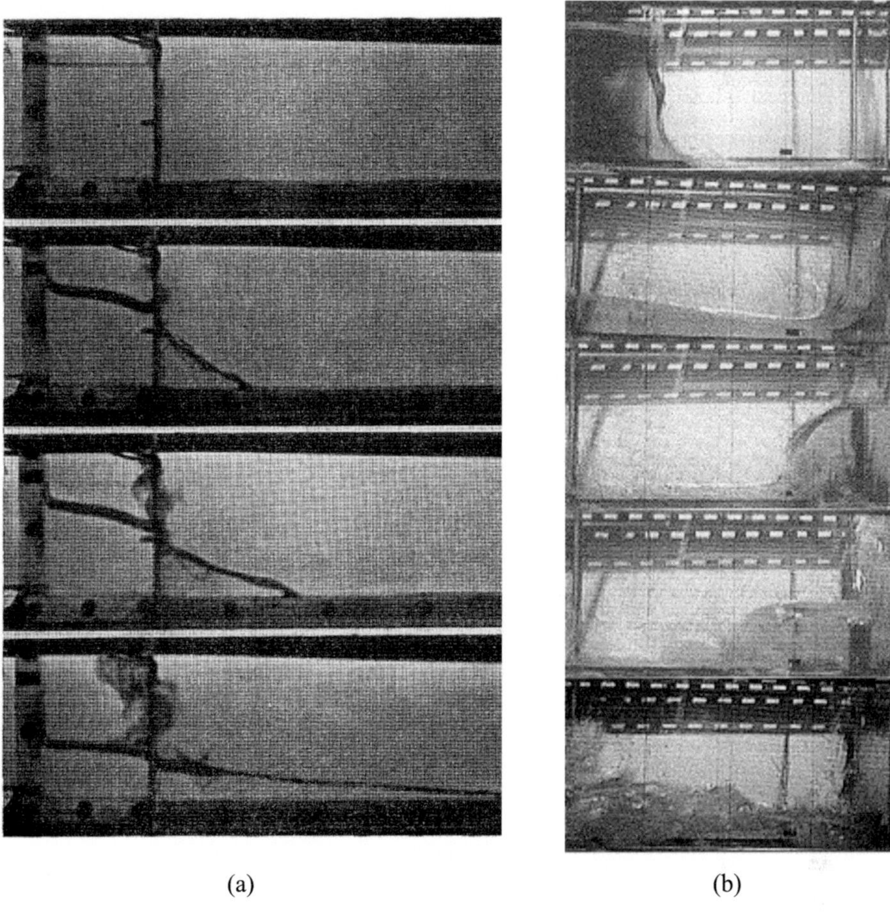

(a)   (b)

Fig. 3.7 Experimental study of the dam break problem:

(a) Martin & Moyce [58] and (b) Koshizuka et al. [47]

The numerical algorithm developed in this study has been applied to the two-dimensional dam break problem. For simplification, a rectangular water column geometry ($n^2 = 1$) has been chosen. The other parameters of the numerical experiments performed are listed in Table 3.1.

Table 3.1. Numerical model parameters for the 2D Dam Break problem

Parameter	Symbol	Unit	Value
Height of fluid column	h_c	m	0.2
Width of fluid column	l	m	0.2
Container width	w	m	0.7
Container height	H_c	m	0.26
Density	ρ_f	kg/m^2	1000
Dynamic viscosity	μ	Pa*s	0.001
Number of particles	N_p	-	63001 (10201)[1]
Interparticle distance	d	m	$8{\cdot}10^{-4}$ $(2{\cdot}10^{-3})$[1]
Smoothing radius	h	m	$8{\cdot}10^{-4}$ $(2{\cdot}10^{-3})$[1]
Distance between fluid and wall particles	d_{fb}	m	$8{\cdot}10^{-4}$ $(2{\cdot}10^{-3})$[1]
Reference constant in EoS	B	kPa	80
XSPH coefficient	ε	-	0.1
Reinitialization period	-	step	20
Wall model	"Fixed" fluid particles		
Viscosity model	Combined model of viscosity (parameter values are set according to Table 2.1)		

[1] Two numerical simulations with different resolutions of the model (number of fluid particles) were performed. The parameters of the experiment with the smaller number of particles are given in the brackets.

The results of the two-dimensional dam break simulation for the case with $N_p = 63001$ are shown in Fig. 3.8. Excessive pressure dP (to the atmospheric pressure) are indicated by different colors.

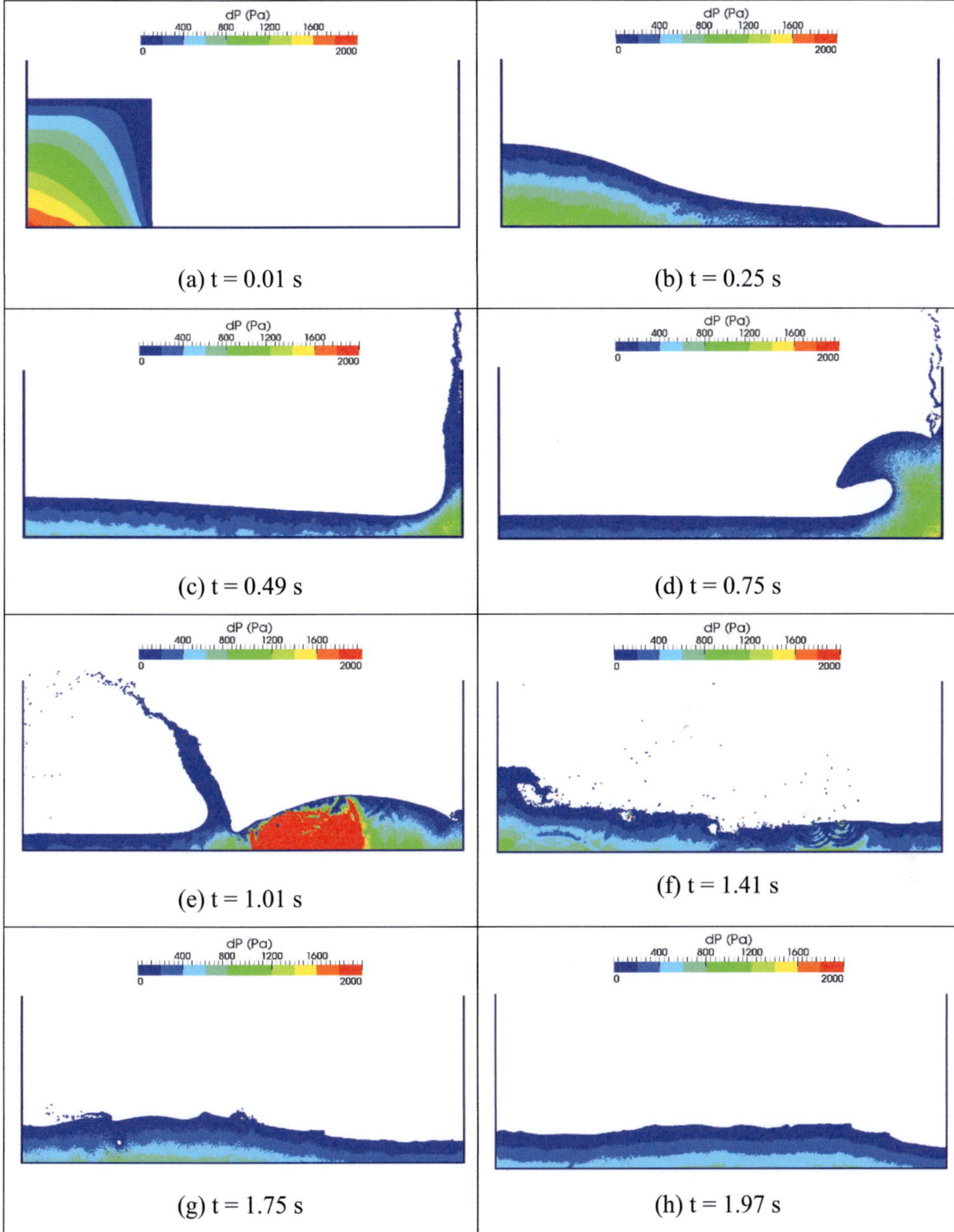

Fig. 3.8. Numerical simulation of the two-dimensional dam break problem: fluid particle distribution ($N_p = 63001$)

Although the available flow snapshots are low quality, it can be seen from Fig. 3.8 that the overall motion of liquid is generally correctly predicted by the numerical algorithm. At the beginning of the simulation, the water column begins to collapse under the force of gravity (a), moves to the right wall (b), and reaches the wall and forms the first slosh (c). The main part of the liquid volume reflects from the vertical wall, while some fragments[1] are separated from the main liquid volume and fly away. The primary backward wave crest is well-represented (d). When the primary backward wave faces the water layer moving toward, the second backward wave is generated (e). At this moment some "air" becomes entrapped under the liquid layer in an oval-shaped void. Later, since no gas phase is simulated, the void collapses, causing strong shock waves in the liquid volume. These effects can also be observed in the experimental results. Further, the wave approaches the left vertical wall (f), forming a new wave crest and a number of droplets. The interactions of these droplets with the water layer result in spherical pressure waves propagating through the water volume. The final state of the system is defined by the travelling secondary waves in the water layer and their interactions with each other and the progressive damping by viscous forces (g-h).

The qualitative comparison of these results with experimental observations allows us to conclude that the present numerical algorithm is generally capable for treating the most important physical effects, such as the formation and propagation of waves along the free surface, their reflection from the walls, and the formation of liquid "drops."

In Fig. 3.8, the result of the numerical simulation with a relatively large number of particles ($N_p = 63001$) is presented. However, for cases with a smaller number of particles, similar behavior of the liquid is also observed overall, but with a rougher shape of the free surface (not shown here).

The quantitative comparison of the predicted results with experimental data has been performed in order to ensure that the current model correctly handles momentum and energy conservation and does not introduce excessive viscosity. The position of the surge front is used as the main quantitative parameter describing the propagation of the wave in the dam break experiment.

---

[1] These fragments look like drops, but since the surface tension model is not implemented in the present algorithm, their sizes and shapes are not physically correct. In the following text, the terms "drops" and "droplets" (in double quotes) are also used to denote these liquid fragments.

In Fig. 3.9, the dimensionless position of the liquid front is plotted versus dimensionless time. The simulation results are compared with the experimental data obtained by Martin & Moyce [58], as well as with the numerical results of the calculations performed with the meshless Moving Particle Semi-implicit (MPS) method by Koshizuka and Oka [47], and with one of the latest developments of the mesh-based Volume-of-Fluid (VOF) technique by Ketabdari et al. [43].

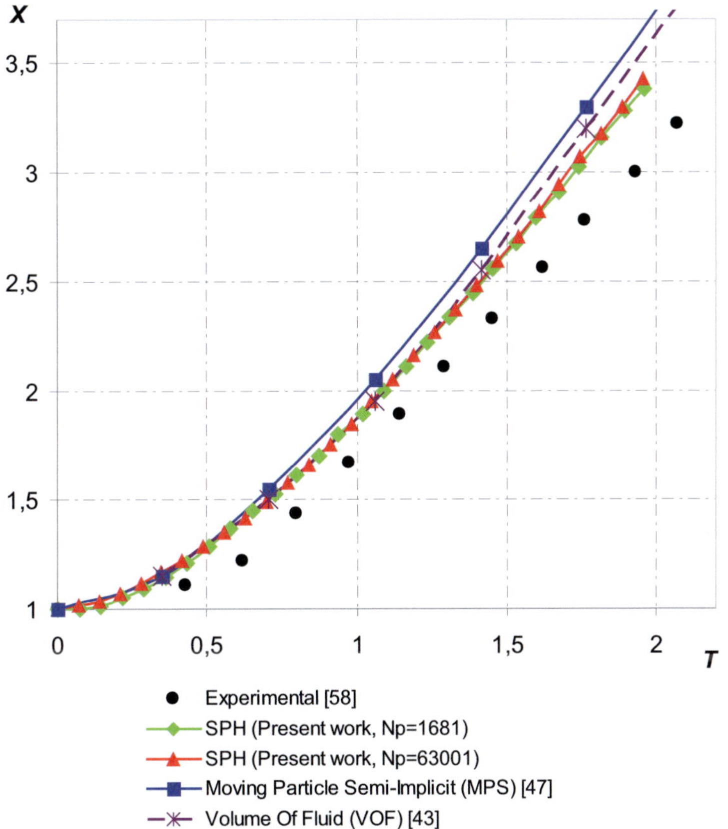

Fig. 3.9. Position of the liquid front in the dam break experiment (dimensionless time

$$T = nt\sqrt{g/l} \text{ and coordinate } X = x/l \text{ are used)}$$

The dimensionless time and position of the wave front are used as coordinates in Fig. 3.9, defined by $T = nt\sqrt{g/l}$ and $X = x/l$, where t, g, l and x are time, standard gravity, initial width of the water column, and position of the wave front. The constant n, in the case of a square water column, is equal to one.

It can be easily seen that all the results of the numerical simulations agree well with each other in the earlier phase. Later in the simulation, the MPS method shows a slightly higher wave front velocity. SPH and VOF simulations predict almost the same front velocity; it is only in the last phase of wave propagation that a difference is observed. At the same time, the experimental data show a slower propagation rate.

There are a number of possible reasons for these discrepancies. The primary reason that they are due to the neglect of friction forces between the liquid and the container walls in all the numerical simulations. In fact, the wall friction effect is partially reproduced by including wall particles in the summation in the viscous term. However, for better predictions, one has to model near-wall flow with a higher number of particles, or else implement semi-empirical models for the wall friction, which was not measured in the available experiments.

Another possible reason is that the numerical experiment does not exactly model the real experimental setup. In the experimental series of Martin & Moyce [58], a thin membrane is used to hold back the water column. At the beginning of the experiment, the membrane is destroyed by an electrical current, but this process is not instant. Taking a look at the experimental snapshots (Fig. 3.7), it can be seen that the membrane is not completely destroyed. The free surface of the liquid volume is not smooth, but has a sharp step break at the point where the membrane is positioned.

Finally, all the presented numerical simulations are performed in two dimensions, although the real experiment is definitely three-dimensional. Some real physical effects, such as spherical (not circular!) drop formation, friction between water and side walls, etc. in principal cannot be considered in the two-dimensional model.

As a final comment, it should be noted that the present simulations with different numbers of particles predict very similar results. The difference never exceeds the measurement error in the position of the surge front.

To estimate the final truncation-converged solution – a solution that represents the case of an infinite number of particles – a technique called Richardson extrapolation [81] has been applied. The procedure is illustrated in Fig. 3.10. The plot shows the dependency of the dimensionless front coordinate X on the initial interparticle distance. Two values of the front coordinate obtained in low resolution ($N_p = 1681$) and high resolution ($N_p = 63001$) runs are plotted and connected with a straight line. These points correspond to the dimensionless time $T = 1.9$. To find an estimation of the truncation-converged solution, the line is extrapolated to the zero initial distance value (dashed slanting line). The X value on the intersection of this line with the vertical axis is assumed to be the solution for an infinite number of particles. The approximation errors for both model resolutions are quantified at 1.0% and 0.2% for the low and high resolution runs, respectively.

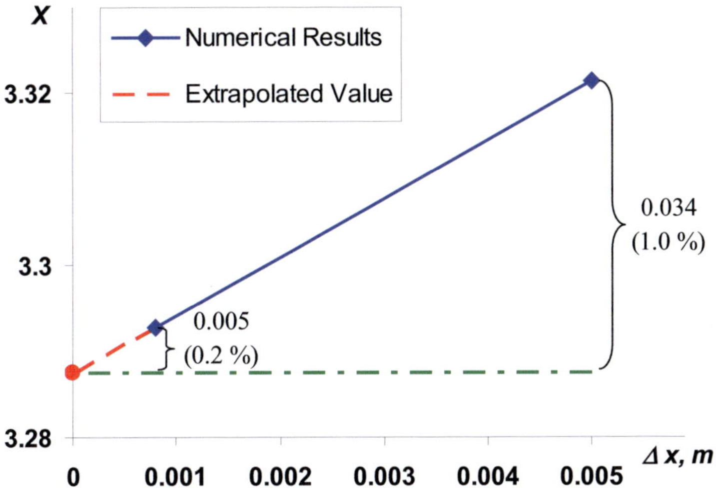

Fig. 3.10. Estimation of approximation errors for two resolutions of the model ($X = x / l$ is a dimensionless coordinate, Δx the initial distance between particles)

One conclusion is that the number of particles required for the simulation of the bulk flow can be fairly small – a mere 1600 particles in two dimensions are sufficient to predict the general flow pattern. However, detailed information about the flow can be obtained only with fine resolution, i.e. with a larger number of particles.

4   Numerical Analysis of the Centralized Liquid Sloshing Experiment

4.1 Introduction

The numerical algorithm described in Section 2, once validated, has been applied to the numerical simulation of the three-dimensional sloshing liquid motion problem. This problem has been experimentally studied in KfK (presently KIT) in the framework of the safety analysis of fast nuclear reactors [59].

The sloshing experiments had two main objectives. The first was to obtain a better understanding of centralized sloshing phenomenon. In a hypothetical severe accident of a fast nuclear reactor, a possible recriticality may occur following core melting and relocation of the fissile materials. In the case of a symmetrical geometry of the damaged reactor core, the relocation of the fuel could happen as a result of the coherent movement of the waves on a free surface of the molten pool. Any disturbances in the coherence or symmetry of the pool would result in a reduction of the compacting volume of the fissile materials, and thus to the reduction of a possible power excursion.

The second purpose of the sloshing liquid motion experiments was to provide data for a benchmark exercise for reactor accident analysis codes [60]. These data were subsequently used to verify and validate the SIMMER-III/IV reactor safety analysis code ([78], [96]).

Obviously, the symmetrical configuration of the sloshing experiment (Fig. 1.3(a)) is most interesting, as it provides perfect conditions for the formation of a critical mass of melted fuel in the center of the core. Snapshots of the corresponding experimental case SA-D1-3 are shown in Fig. 4.1.

Fig. 4.1. Central sloshing experiment. Experimental series SA-D1-3 (symmetrical geometry) [60]

In addition to the SA experimental series, experimental investigations of various "non-ideal" geometries were performed to prove that small deviations in symmetry can dramatically reduce the possibility of a recriticality event. These included geometries with asymmetrical locations of the fluid column without obstacles on the flow area (series SB), as well as symmetrically releasing liquid columns surrounded by obstacles, such as rod bank imitators (series SD) or ring walls (series SC). The impact of small particles on flow behavior was also studied in the SE experimental series. Snapshots from the experimental case SB-D1A-2 of the sloshing experiments are shown in Fig. 4.2.

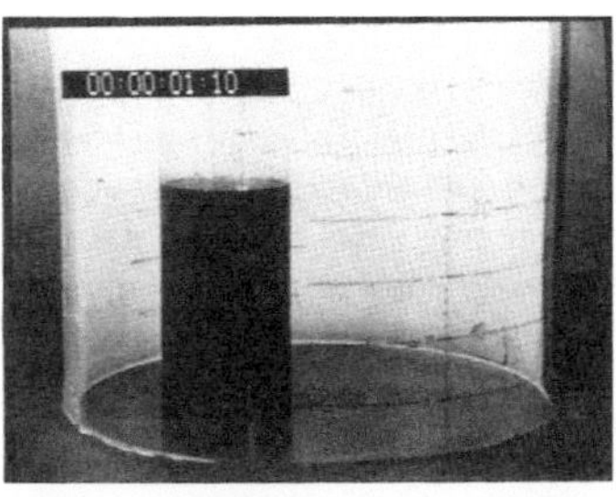

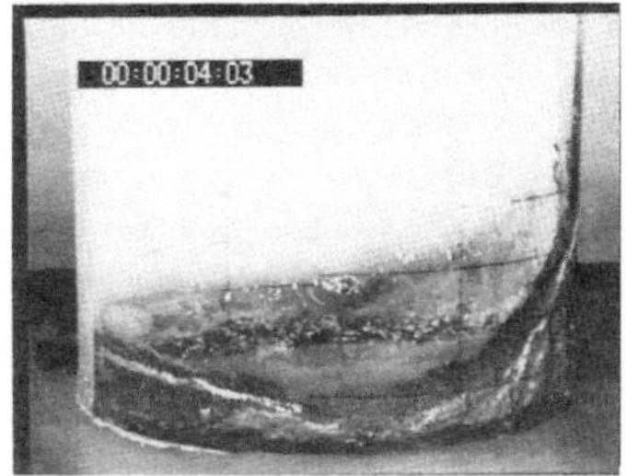
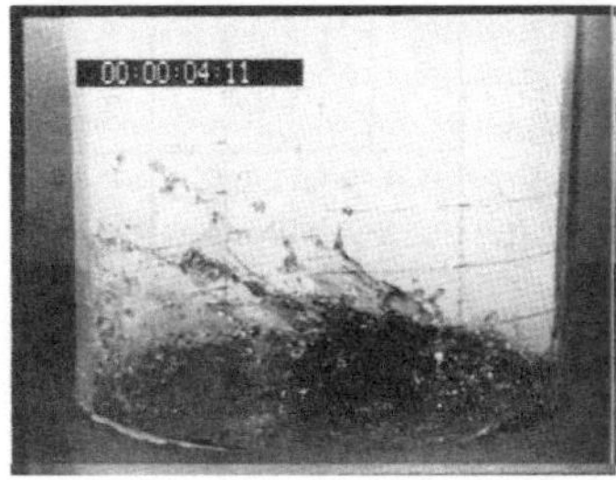

Fig. 4.2. Central sloshing experiment. Experimental series SB-D1A-2 (asymmetrical geometry) [60]

This chapter deals with the numerical simulation of the following experimental configurations:

- a fully symmetrical case;
- an asymmetrical case with no obstacles in the flow;
- a symmetrical case with a rod bank installed around the liquid column.

The results of the numerical simulations are compared with the experimental data, as well as with the results of the numerical simulations obtained from the SIMMER-III/IV code. The numerical simulations presented in this section are also described in [109] and [110].

4.2 Simulation of Experimental Series without Flow Obstacles

4.2.1 Geometry of Computational Domain and Calculation Parameters

It is clear that the perfect symmetrical geometry is the most dangerous configuration for possible recriticality. A series of numerical calculations have been performed for the experimental case SA-D1-3. To estimate the possibility of recriticality in this case, an evaluation of the liquid masses compacting at the center of the pool was performed in [60]. It was found that, at the time when the maximal central peak is observed, approximately 34% of the liquid is located in the area originally occupied by the liquid column (the cylindrical area in the container center with 11 cm diameter). The mass of the lower third of the central peak was estimated at 70% of the liquid located in the central area.

The geometry of the computational domain used in numerical simulations is shown in Fig. 4.3.

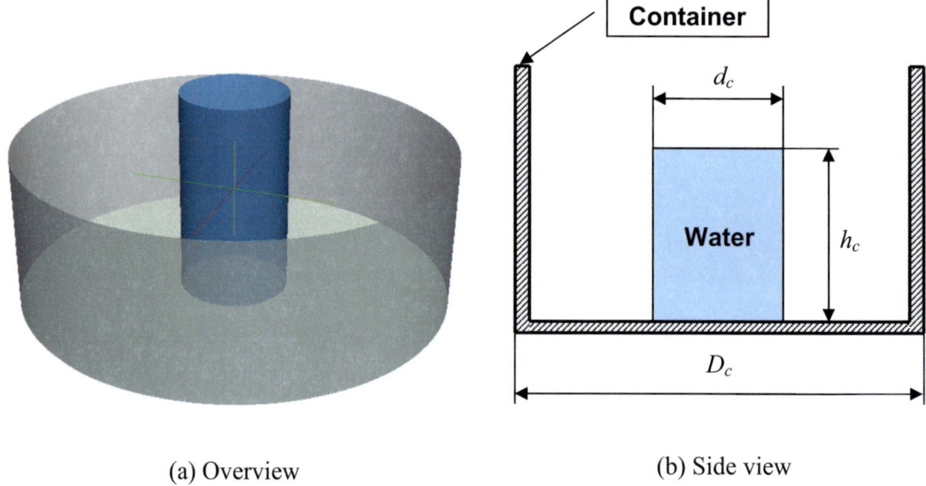

(a) Overview   (b) Side view

Fig. 4.3. 3D central sloshing: geometry of the computational domain

The parameters of the numerical model used in the calculations for the symmetrical geometry are listed in Table 4.1.

Table 4.1. Parameters of the numerical model: 3D Central Sloshing

Parameter	Symbol	Unit	Value
Height of water column	h_c	m	0.2
Diameter of water column	d_c	m	0.11
Container diameter	D_c	m	0.44
Height of container walls	H_c	m	0.20
Fluid density	ρ_f	kg/m^3	1000

Dynamic viscosity	μ	Pa*s	0.001
Number of particles	N_p	-	71422 (189313[1])
Interparticle distance	d	m	$3 \cdot 10^{-4}$
Smoothing radius	h	m	$3 \cdot 10^{-4}$
Distance between fluid and wall particles	d_{fb}	m	$3 \cdot 10^{-4}$
Reference constant in EoS	B	kPa	60
XSPH coefficient	ε	-	0.1
Wall model	"Fixed" fluid particles		
Viscosity model	Combined model of viscosity (parameter values are set according to Table 2.1)		

For the asymmetrical case, the same computational domain geometry is used, but the position of the water column is shifted by an offset of 8.25 cm from the container center. The corresponding experimental case is SB-D1A-2 [60].

4.2.2 Results and Discussion

In this subsection, the results of the simulation of the centralized sloshing experiment in both the symmetrical geometry (experimental case SA-D1-3) and the asymmetrical geometry (experimental case SB-D1A-2) are presented. Fig. 4.4 shows a visualization of the results of the simulation in comparison with the experimental observations of the liquid sloshing motion. The vertical component of the particle velocities are indicated in color. The time stamps located in the left top corner of the experimental snapshots do not correspond to the real time of the experiment. The timings below the pictures are taken from the numerical simulations. The same indications of time and velocity are used later, in Fig. 4.6, Fig. 4.12, and Fig. 4.13.

---

[1] The number of wall particles.

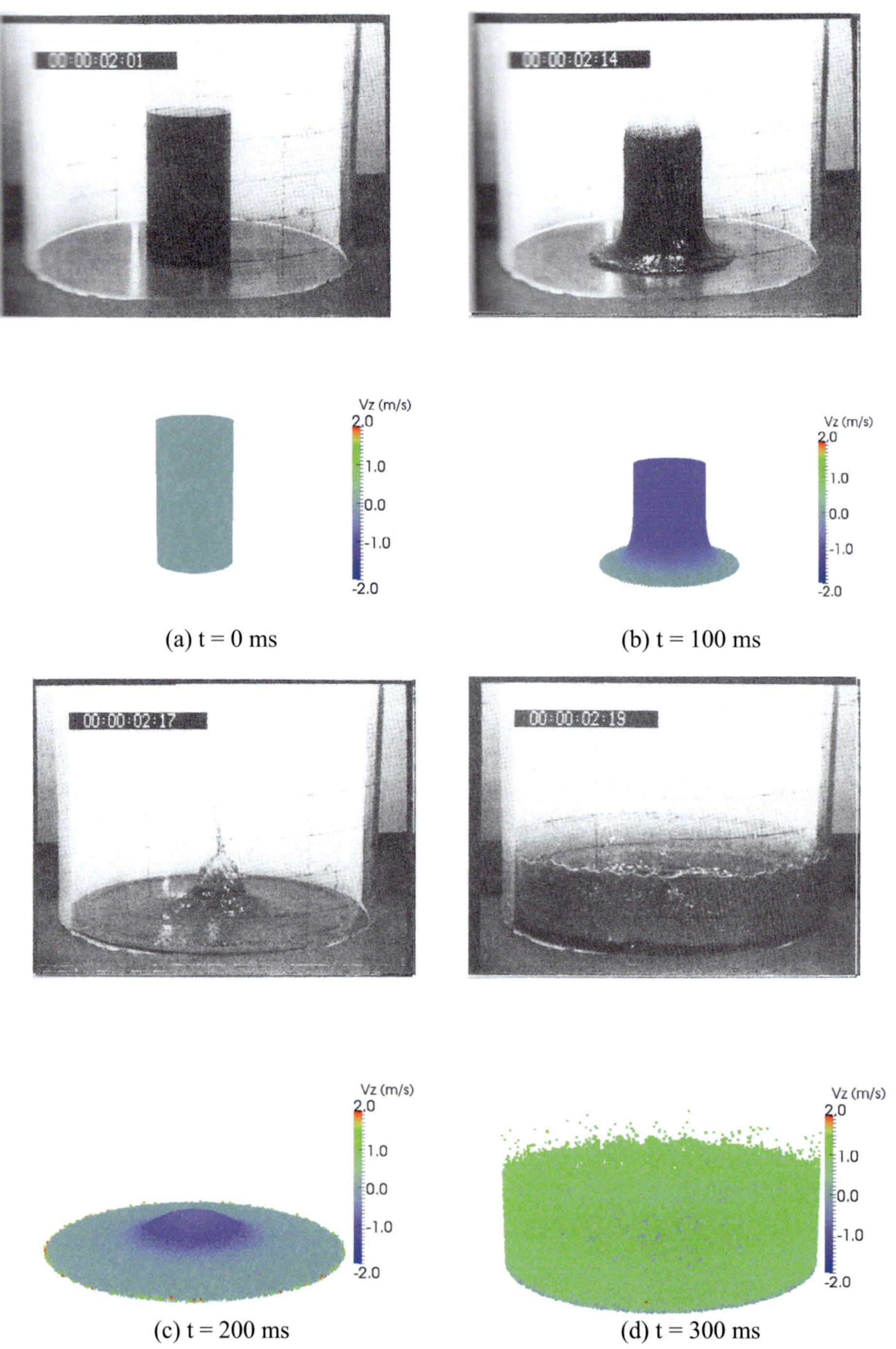

(a) t = 0 ms

(b) t = 100 ms

(c) t = 200 ms

(d) t = 300 ms

Fig. 4.4. 3D central sloshing: case SA-D1-3 (symmetrical)

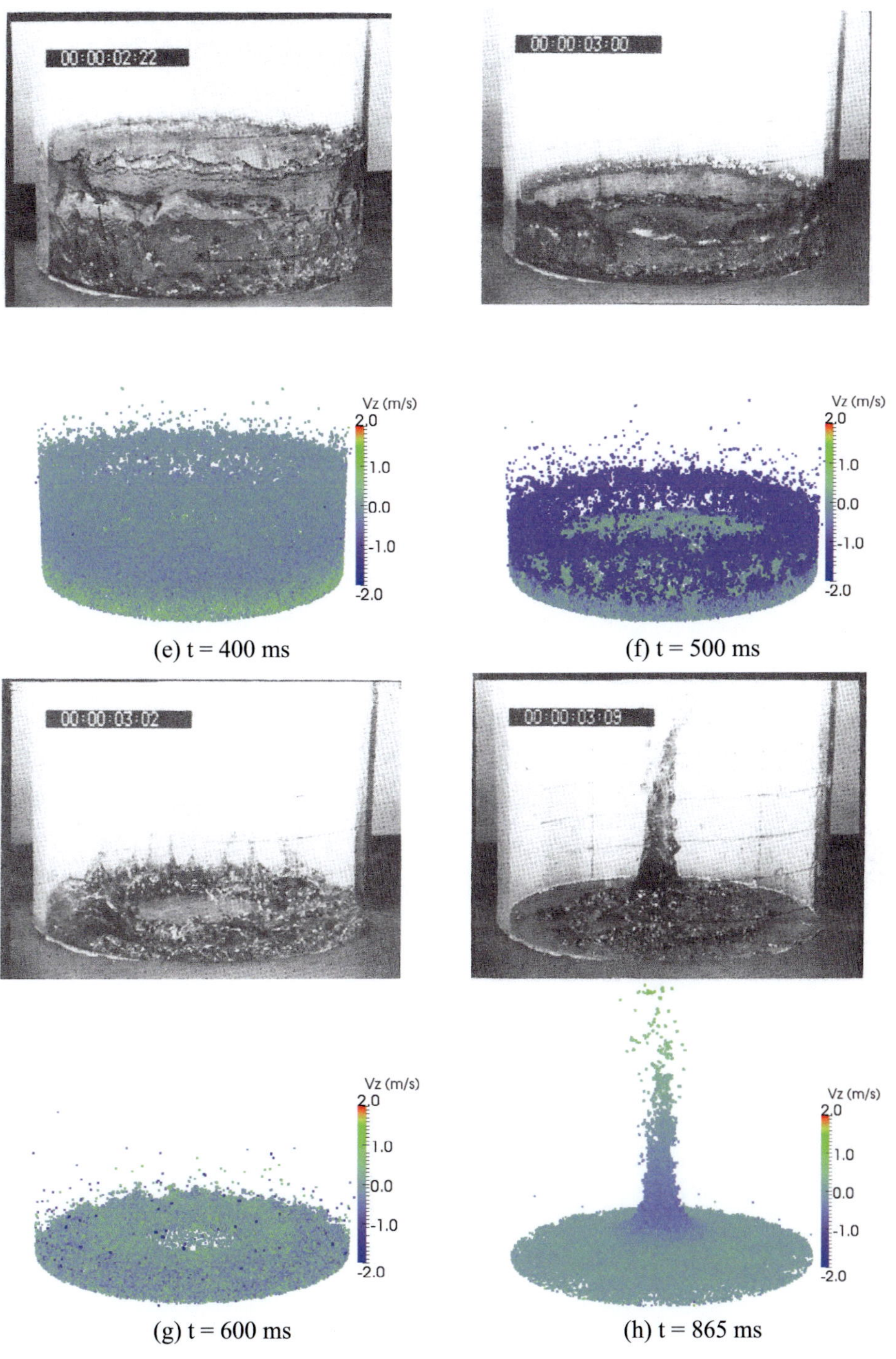

(e) t = 400 ms

(f) t = 500 ms

(g) t = 600 ms

(h) t = 865 ms

Fig. 4.4 (continued). 3D central sloshing: case SA-D1-3 (symmetrical)

At the initial moment (Fig. 4.4(a)), the liquid column is released and, under the influence of gravitation forces, starts to collapse. The collapsing column creates a circular wave which moves towards the container wall (Fig. 4.4(b), and having touched the wall (Fig. 4.4(c)), forms the outward slosh (Fig. 4.4(d-e)). The slosh at the container wall loses its kinetic energy, and reaching the maximum, begins to collapse. This results in the generation of a converging wave approaching the center of the container (Fig. 4.4(f-g)). Later, liquid moving in the direction of the container center forms a high central peak (Fig. 4.4(h)). At the edge of the peak, several liquid fragments are separated from the bulk volume and fly away. Other fragments follow the main volume of liquid and progressively lose their velocity. When $t \sim 860$ ms, the central peak with its surrounding "droplets" reaches its maximum height and begins to collapse. Further system states are defined by the motion of the secondary waves and their interaction with the container wall and with each other. Such interactions result in several secondary sloshes at the wall and at the center of the container. Finally, the sloshing motion is completely damped out.

Comparing the numerical results with the experimental snapshots, it can be seen that the overall flow pattern is well represented in the numerical model. The SPH algorithm predicts the evolution of the liquid system with good resolution of the free surface effects. Fine conservation of the total volume of liquid is provided by the correct choice of the "numerical" speed of sound (as described in Section 2.2).

For illustration purposes, the visualized simulation results obtained by Yamano et al. [117] using the SIMMER-IV code for the same experimental case are presented in Fig. 4.5.

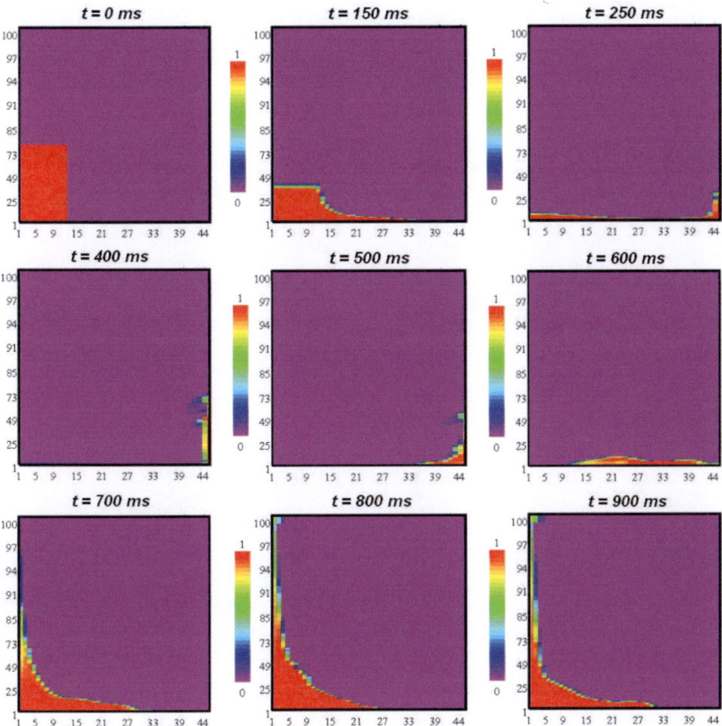

Fig. 4.5. Simulation of the 3D Central Sloshing experiment (symmetrical case) with the
SIMMER-IV code [117]

The results obtained from SIMMER-IV show less information about the flow shape than do the results of the simulations with the SPH method. SIMMER-IV does not reproduce the "droplet" formation on the free surface (these "droplets" can be seen very clearly in the snapshots of the experiment), while the SPH method does. The significant conclusion is that the meshless nature of the SPH method allows fragmentation of the liquid (the formation of small "drops") to be easily predicted, which is a severe challenge for traditional mesh-based CFD methods.

For the asymmetrical case (SB-D1A-2), a visualization of the simulation results is shown in Fig. 4.6, where it is compared with the experimental observation of the liquid sloshing motion. The velocity and time indications correspond to those used in Fig. 4.4 (see comments on page 61).

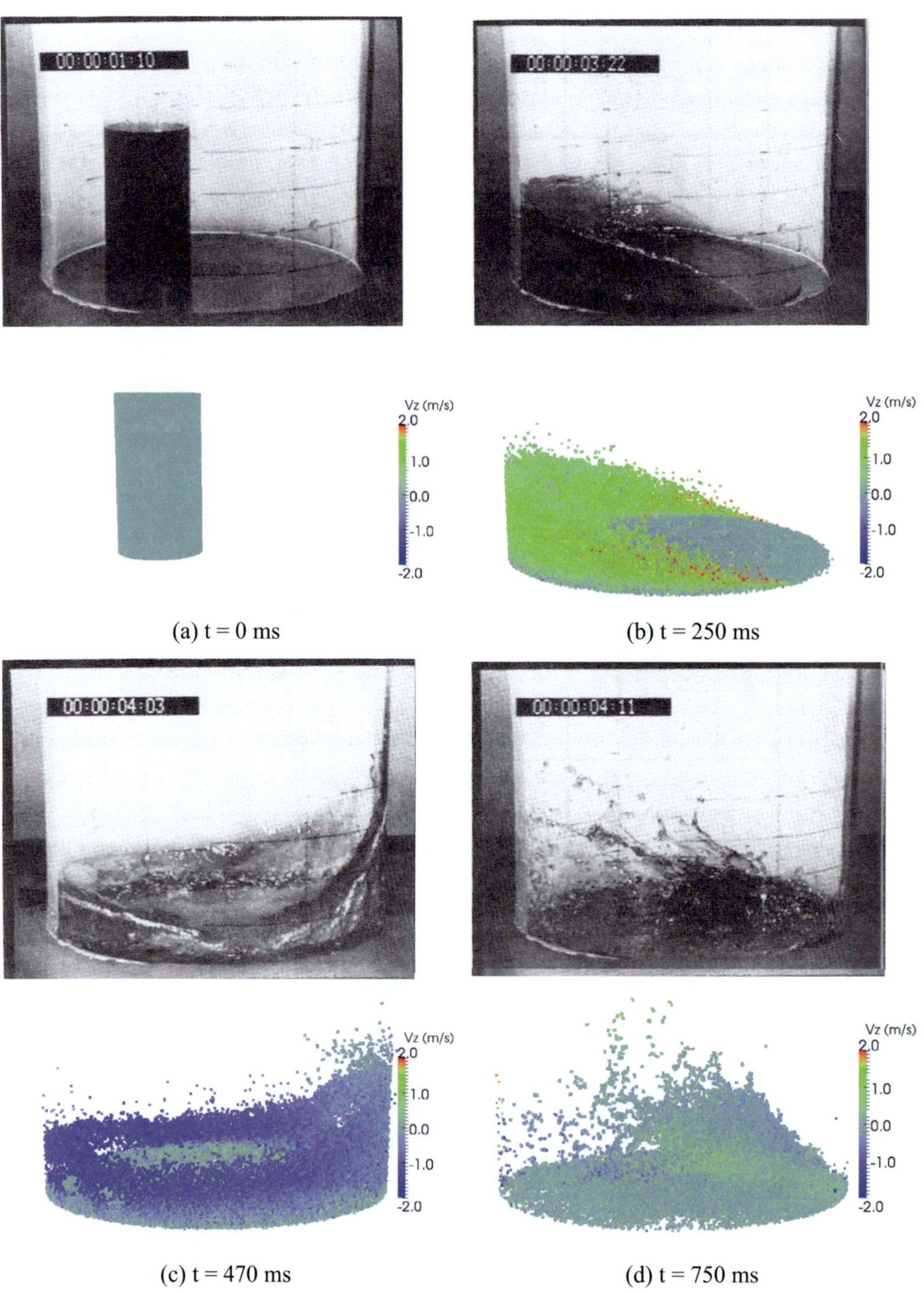

(a) t = 0 ms

(b) t = 250 ms

(c) t = 470 ms

(d) t = 750 ms

Fig. 4.6. 3D central sloshing: case SB-D1A-2 (asymmetrical)

Similarly to the symmetrical test case, the liquid column begins to collapse after release (Fig. 4.6(a)). However, due to the significant offset of the initial column position from the center, it first approaches the left (closest) wall of the container and forms the first slosh there (Fig. 4.6(b)). When the slosh reaches its maximum, the primary wave generated by the collapsing column of liquid approaches the right wall, forming the secondary slosh at that wall (Fig. 4.6(c)). Strictly speaking, the primary wave results in one circular slosh along the container wall, with the first maximum at the left container point and the last maximum at the right container point. After reflection from the wall, the secondary waves meet in between the container center and its right wall. They form a highly distorted peak with a complicated free surface shape (Fig. 4.6(d)). A large number of "droplets" are also generated. Later, the peak collapses in the container, resulting in further oscillations of the free surface of the flow.

Comparing the visualized results with the experimental snapshots presented in [60], one can conclude that the numerical results qualitatively repeat the processes which take place in the experimental series. The general shape of the flow is well represented, although the individual shapes and positions of the "droplets" can differ from those experimentally observed. It is important to note that the real flow is highly distorted and the particular configuration of the free surface of the flow also differs in repeating experiments [60].

In order to make a quantitative assessment, the experimental data [60] are compared with the present numerical simulations, as well as with the numerical results calculated by Yamano et al. [117] using the 3D reactor code SIMMER-IV.

The main quantitative parameters for the symmetrical case are the arrival time of the liquid at the wall, the time and height of the maximal wave at the wall, and the time and height of the central peak. The central peak height is the most important of these for the recriticality analysis, and, as has been found, the peak height is difficult to reproduce in the numerical simulation. Furthermore, a correct definition and measurement of the central peak height is not very obvious. In the applied experimental technique, large drops on top of the peak moving with the same velocity as the bulk flow were included in the height measurement (see the more detailed discussion of the definition of the central peak value in the following subsection). For the asymmetrical case, only the timing of the maximum height at the walls and the height of the maxima were measured in the experiments.

The results are summarized in Table 4.2 (symmetrical case, SA-D1-3) and Table 4.3 (asymmetrical case, SB-D1A-2).

Table 4.2. 3D Central Sloshing Simulation. Symmetrical Case

SA-D1-3	Slosh at outer container wall			Slosh at pool center	
	Arrival time at wall [s]	Time of maximum height [s]	Maximum height [cm]	Time of maximum height [s]	Maximum height [cm]
Experiment	0.20 ± 0.02	0.42 ± 0.02	16.0 ± 1.0	0.88 ± 0.04	40.0 ± 5
Present work (low resolution, $\sim71 \times 10^3$ part.)	0.20 ± 0.02	0.39 ± 0.02	16.5 ± 1.1	0.85 ± 0.05	32.5 ± 11.5
Present work (high resolution, $\sim128 \times 10^3$ part.)	0.19 ± 0.01	0.40 ± 0.01	17.0 ± 0.6	0.87 ± 0.03	38.0 ± 6.0
SIMMER-IV (coarse mesh - 44×44×100)	0.20	0.4	17.25	0.88	36.0
SIMMER-IV (fine mesh - 92×92×100)	0.20	0.38	18.75	-	>50 (overestimated)

Table 4.3. 3D Central Sloshing Simulation. Asymmetrical Case

SB-D1A-2	Slosh at left wall		Slosh at right wall	
	Time of maximum height at left wall [s]	Maximum height (left) [cm]	Time of maximum height at right wall [s]	Maximum height (right) [cm]
Experiment	0.36 ± 0.02	14.0 ± 2.0	0.48 ± 0.02	24.0 ± 2.0
Present work (low resolution)	$0.34 \pm 0.02^{*}$	$15.5 \pm 1.0^{*}$	$0.47 \pm 0.03^{*}$	$20.5 \pm 1.3^{*}$
SIMMER-IV	0.36	17.25	0.48	21.0

[*] The error range is defined using the percentage error values from the simulation of SA-D1-3 experimental case (see Table 4.2)

Most of the quantitative simulation results are in a good agreement with both experimental data and the numerical results predicted by the reactor safety analysis code SIMMER-IV, although some deviations in the central peak value are observed. Also, for the asymmetrical case, a lower value of the height of the right-hand slosh is predicted. A similar value was obtained using the SIMMER-IV code. For the asymmetrical geometry, the lower values may be due to the relatively low resolution of the numerical model.

To estimate the truncation error, a Richardson extrapolation technique [81] has been applied. The procedure is similar to the one used in Section 3.3 for the estimation of the truncation error in the position of the wave front in the dam break simulation. An illustration of the technique is given in Fig. 4.7, where the peak height values for different model resolutions are plotted. Supposing a linear dependence between the central peak height value and the $1/N_p$, where N_p is the total number of fluid particles, a 44 cm central peak height value can be obtained as an estimate for an infinite number of particles.

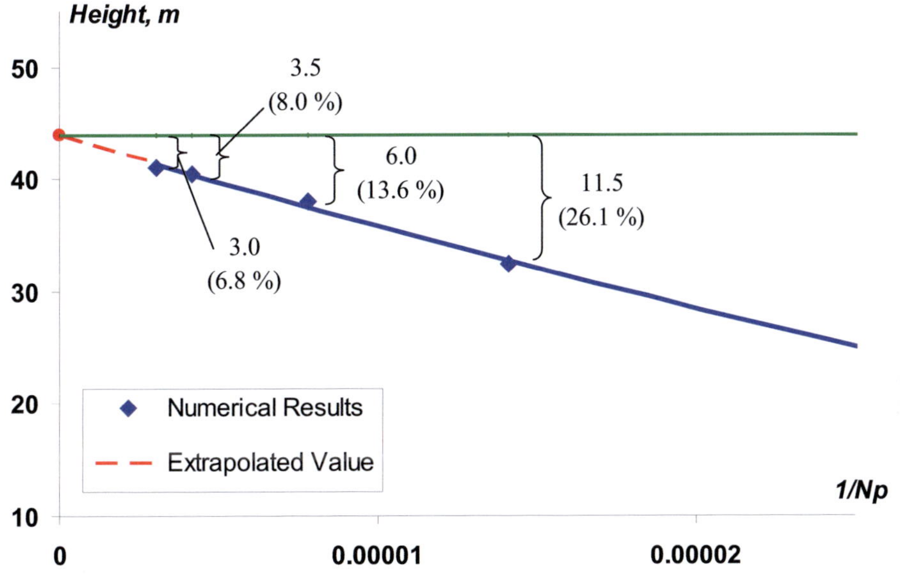

Fig. 4.7. Estimation of the approximation errors in the central peak height value (N_p is the total number of fluid particles representing the liquid)

In the same way, the truncation errors for other quantitative parameters, listed in Table 4.2, were estimated. The percentage error values, obtained for the symmetric sloshing simulation results, were also used to obtain absolute error values for the simulation results of the series SB-D1A-2 (asymmetrical case) and SD-D1R-1(2) (cases with rods).

4.2.3 Central Peak Height Value for the Symmetrical Test Case. Sensitivity Study

The height of the central sloshing peak can be defined in several ways. The question here is which drops at the top should be included as a part of the peak, and which should be excluded. In the present numerical analysis, the vertical component of velocity is examined. At the moment when the peak height reaches its maximum and begins to collapse, the "drops" with a negative vertical component of velocity are included in the definition of the height of the peak, while those "drops" with a positive vertical velocity are not. In SIMMER-IV, the height of peak is defined as the maximal location of cells with a liquid fraction, because liquid drops are not simulated. The definition of the peak height in the experiments is rather subjective, and does not include small drops above the central peak.

The present simulations with lower resolution (about 71×10^3 particles) underestimate the height of the central peak (32.5 cm) compared to the experimental value (40.0 ± 5 cm). SIMMER-IV also predicts a smaller value for the central peak height (38 cm), but the difference is within the experimental measurement error range.

The numerical results obtained from the low resolution simulation are affected by the leakage of some particles above the peak of the bulk liquid, which results in a smaller value of the peak diameter compared with experimental observations. In addition, the top part of the central peak is more fragmented than the liquid in the experiment. To overcome the leakage of particles, which can affect the results, another series of numerical simulations for the symmetrical case SA-D1-3 was performed with higher resolution (about $\sim128 \times 10^3$ particles). The geometrical and physical parameters of the numerical model were in line with Table 4.1, excluding the initial distance between particles d, the smoothing radius h, and the initial distance between fluid and wall particles d_{fb}, which were set to 0.25.

The high-resolution simulation predicts a higher value of the peak height (38 cm), which is slightly lower than the experimental value, but which lies within the error range of the measurements. It should also be noted that an experimental value of 40 cm is the maximal value obtained in the particular test SA-D1-3, which was one of a series of several tests. In many other experimental cases, the central peak was not as high, or was even unobserved, due to small deviations from "ideal" symmetry conditions.

In the high-resolution simulations, the shape of the peak became smoother compared to the low-resolution runs, and also agreed better with the actual peak observed in the experiment.

Particularly, the volume of liquid involved in the peak formation was significantly larger, although the value of the peak height was only slightly higher compared with the low-resolution simulation. To illustrate this difference between the low and high resolution simulations, two-dimensional projections of the particle system at the instants of maximum height are shown in Fig. 4.8 (a) and Fig. 4.8 (b), respectively.

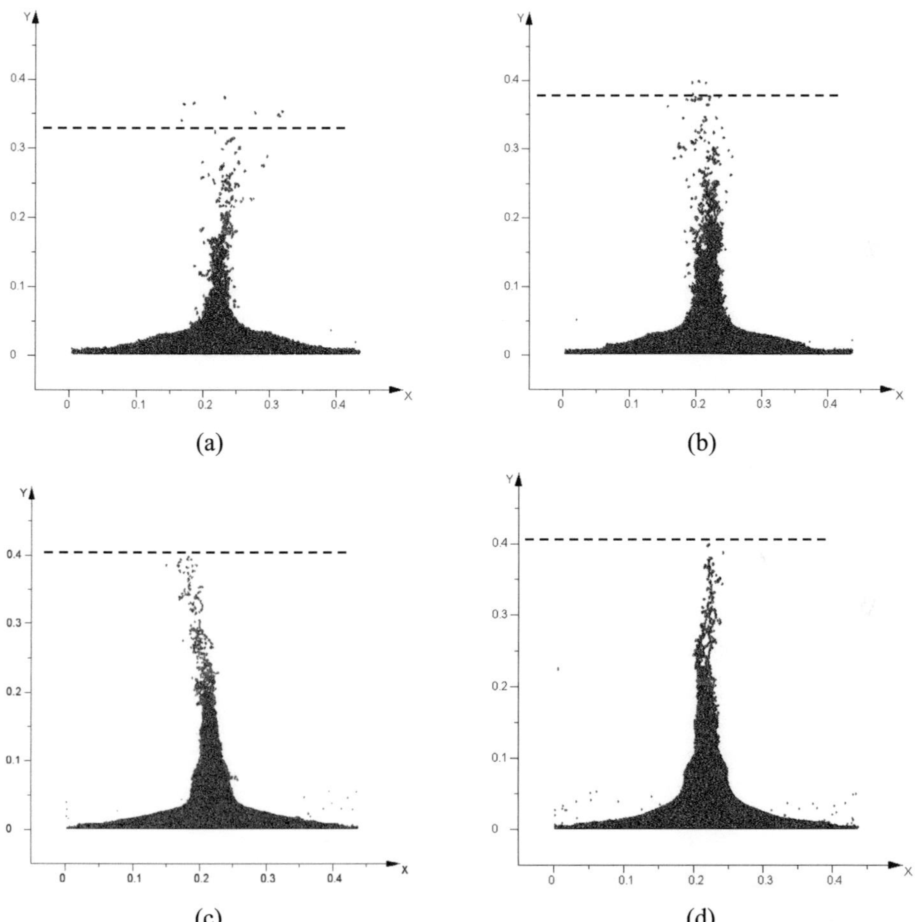

Fig. 4.8. 3D central sloshing. Comparative view of the central peak in the symmetrical case (lengths along the axes in [m]). The dashed lines show estimated values of the peak height, based on the vertical component of velocity. (a) Number of particles representing fluid column $Np \sim 71 \times 10^3$; (b) $Np \sim 128 \times 10^3$; (c) $Np \sim 238 \times 10^3$; (d) $Np \sim 326 \times 10^3$;

As a sensitivity study, two additional numerical simulations with $\sim 238 \times 10^3$ and $\sim 326 \times 10^3$ fluid particles representing the fluid column were performed. The shape of the peak is shown in Fig. 4.8 (c, d). Several liquid fragments are found at a higher position, similar to the cases with lower resolutions. Such small, fast moving fragments are not taken into account. In Fig. 4.9, the comparative heights of the central peak versus the number of the fluid particles used for the simulation are shown.

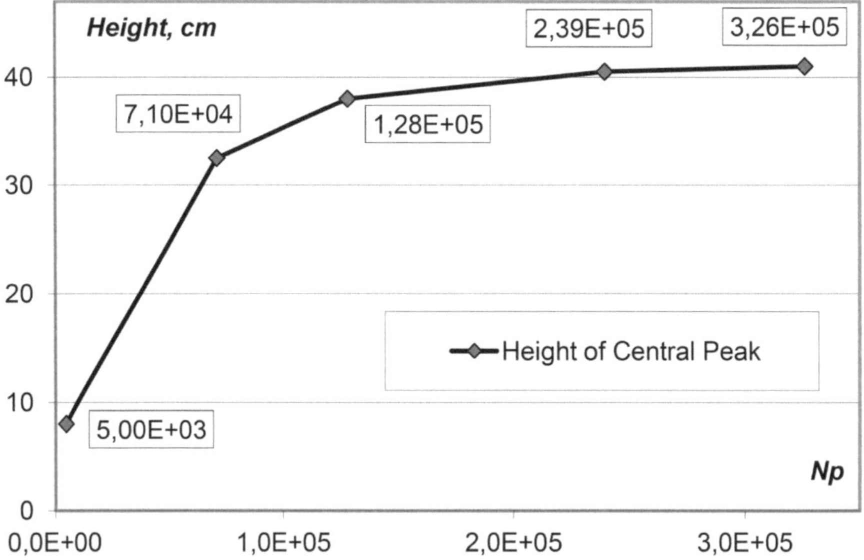

Fig. 4.9. Heights of the central peak obtained with different resolutions of the numerical model

The calculated value of the central peak height tends to a certain upper bound higher than the experimental value of 40 cm. However, when increasing the number of particles, it should be mentioned that the definition of the height cannot be made precisely, whether in the numerical calculations or in the experiment, due to the existence of separate splashes above the liquid column in both. These estimations are mainly made based on visual observations.

Another estimation of the peak height can be made using particle distribution data. The idea is to take the height below which 95% of fluid particles are located as the maximal peak height. At the moment when the maximal central peak is observed, the lower part (from the bottom to the 10 cm level) of the vessel contains 97% of the fluid particles. For this reason, an analysis of the particle distribution along the vertical axis was performed only for those particles positioned higher than 20 cm from the bottom of the container.

Fig. 4.10 plots the number of particles located in the interval $[z; z + 1$ cm] against the vertical coordinate z (height), using the data distribution from the simulation of the symmetrical sloshing experiment with 326×10^3 fluid particles. It can be seen from the data plot that the number of particles located at a specific height significantly decreases from the 20 cm level to the 35 cm level. At the 36 cm level and above, the number of particles is reduced by an order of two compared with the value at the 20 cm level. Summing the number of particles layer by layer up to 95% of all fluid particles, the peak height is estimated as 37 cm. This value is indicated on the plot as a vertical dashed line.

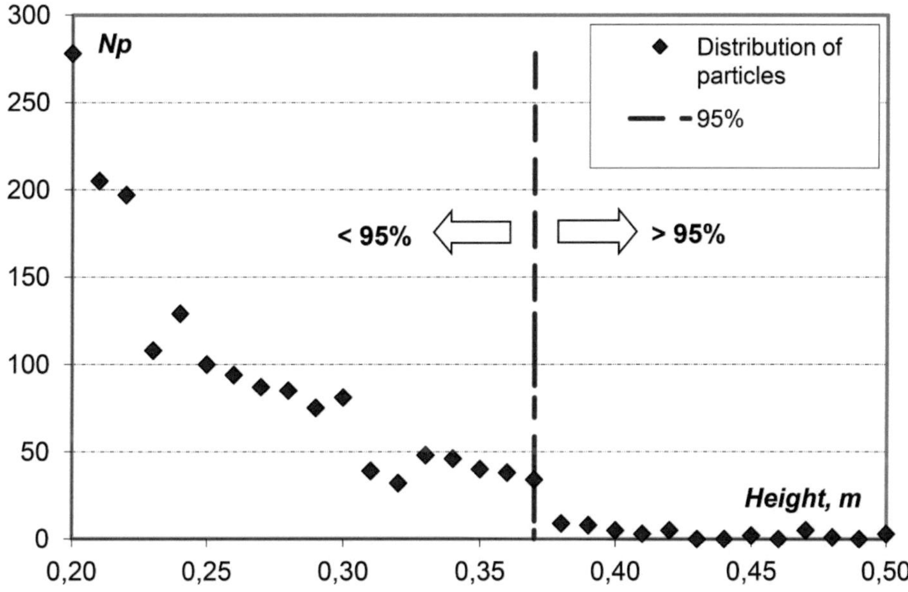

Fig. 4.10. Distribution of the fluid particles along the vertical coordinate

A notable observation is that SIMMER-IV overestimates the central peak height with an increase of the mesh resolution. This means that, for the highest resolution case, the central peak is even larger than the computational domain, and there is a priori no possibility of defining the height and time of the maximum. This problem is probably related to the strong mesh symmetry in the SIMMER-IV simulations. A hypothetical symmetric solution should also significantly overestimate the central peak height, due to the perfect cylindrical symmetry. In reality, there is no perfect symmetry – small deviations which disturb symmetry always occur, resulting in a reduction of the central peak height. From this point of view, particle methods are more suitable for simulations because they represent the behavior of a system by the movement of a large number of particles, and are therefore closer to the real and not perfectly symmetrical conditions.

Another part of the problem is that the computational domain in mesh-based methods is limited by the size of the mesh, and it is not possible to get any information about processes which take place outside of the domain, unless the mesh is expanded up to the maximum possible drop position. The SPH method, like other meshless methods, is free from this limitation – all particles can be tracked as far as necessary.

4.3 Simulation of the Experimental Series with a Rod Bank

4.3.1 Geometry and Set-up

The geometry of the numerical model for the test cases with rod imitators is the same as for the experimental series without rods, described in Section 4.2. The difference is the presence of twelve vertical rods equidistantly positioned around the water column. Their distance from the center is $R_c = 9.9$ cm or $R_c = 17.6$ cm respectively for the experimental series SD-D1R-1 and SD-D1R-2. The rod diameter (d_{rod}) in the experiments was 2 cm, to simulate a blockage ratio similar to that in a real reactor pool. The same value for the rod diameter has been used in the numerical model. An overview and sketch of the experimental setup, with geometrical sizes, for these test series are given in Fig. 4.11.

Both test cases were simulated using $\sim 128 \times 10^3$ fluid particles to represent the water column and $\sim 371 \times 10^3$ "fixed" fluid particles for the walls (including the particles used for rod modeling).

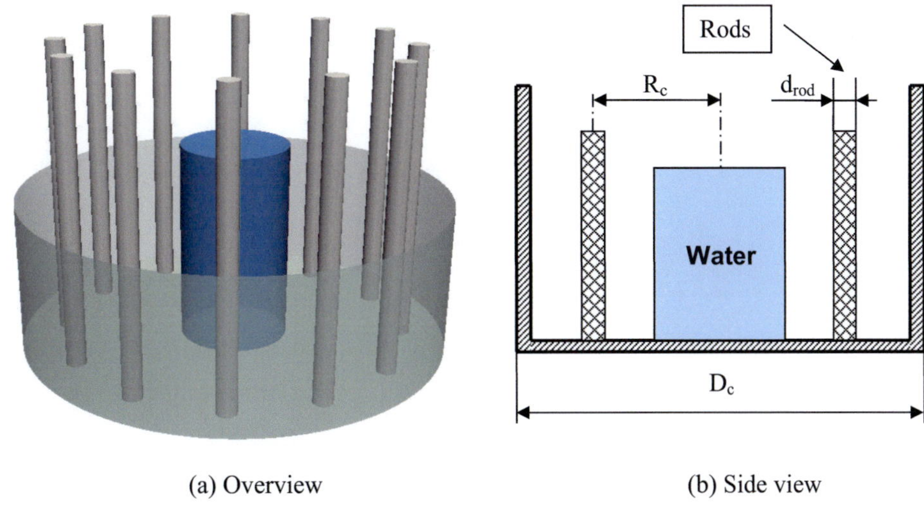

(a) Overview   (b) Side view

Fig. 4.11. 3D central sloshing: experimental series with the rod bank. Geometry of the computational domain

4.3.2 Simulation Results

The comparative snapshots from the experiment and numerical simulation for the test cases SD-D1R-1 and SD-D1R-2 are presented in Fig. 4.12 and Fig. 4.13, respectively. In order to get a better view of the flow evolution, the container walls and the rods are not shown. The velocity and time indications correspond to those used in Fig. 4.4 (see comments on page 61).

The quantitative values for comparison, which are similar to the symmetrical sloshing simulation, are presented in Table 4.4 and Table 4.5 for tests SD-D1R-1 and SD-D1R-2, respectively. No numerical simulation results for test series SD (flow with rods) are available in the literature, so the simulation results are compared with the experimental data only. For each geometry, three numerical simulations have been performed with different parameters for the viscosity models:

    – Artificial viscosity model with $\alpha = 0.01$, $\beta = 0.1$;

    – Combined viscosity model with $\mu = 0.001$ Pa·s, $\beta = 0.1$;

    – Artificial viscosity model with $\alpha = 0.01$, $\beta = 0.1$ and extra wall friction (see Section 4.3.3.2 for details).

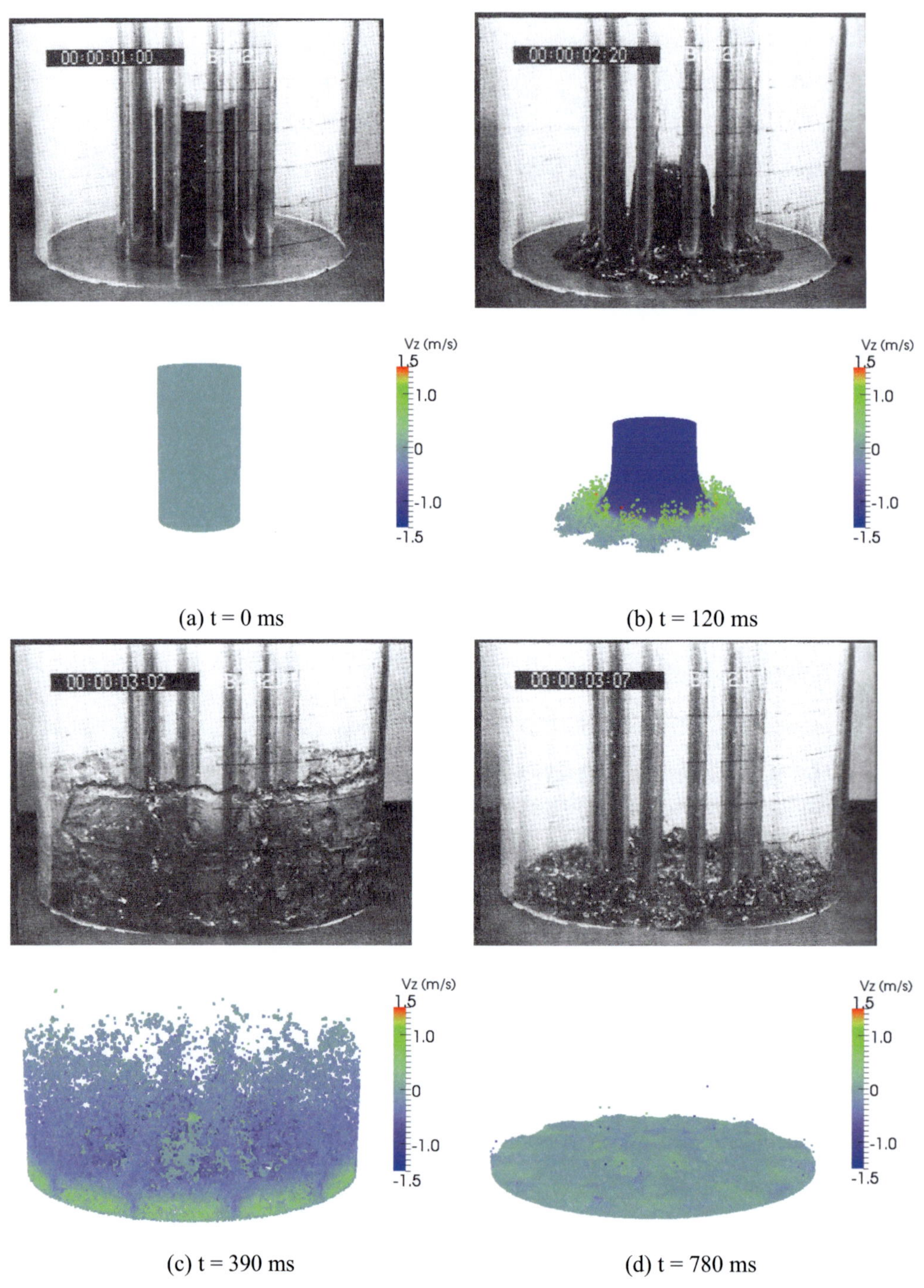

(a) t = 0 ms

(b) t = 120 ms

(c) t = 390 ms

(d) t = 780 ms

Fig. 4.12. 3D central sloshing: case SD-D1R-1 (with rods, $R_c = 9.9$ cm)

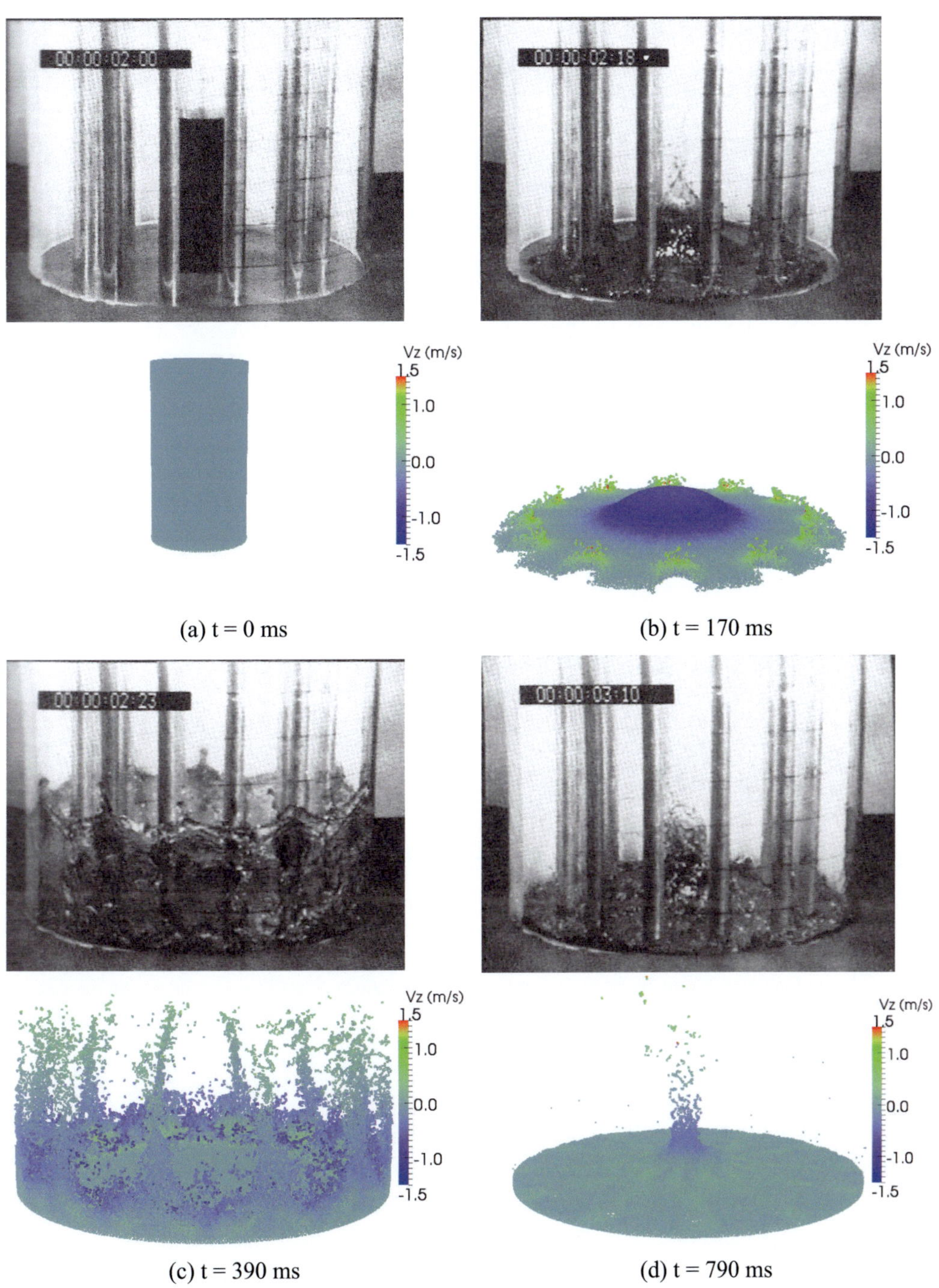

(a) t = 0 ms (b) t = 170 ms

(c) t = 390 ms (d) t = 790 ms

Fig. 4.13. 3D central sloshing: case SD-D1R-2 (with rods, R_c = 17.6 cm)

Table 4.4. Sloshing Experiment: Vertical Rod Bank (Test Case SD-D1R-1)

SD-D1R-1 (Dr = 9.9 cm)	Slosh at outer wall			Slosh at pool center	
	Arrival time at wall [s]	Time of maximum height [s]	Maximum height [cm]	Time of maximum height [s]	Maximum height [cm]
Experiment	0.22 ± 0.02	0.44 ± 0.02	15 ± 1	0.90 ± 0.04	3 ± 2
Artificial viscosity	$0.19 \pm 0.01^*$	$0.39 \pm 0.01^*$	$18.5 \pm 0.6^*$	$0.78 \pm 0.02^*$	$4.5 \pm 0.6^*$
Combined viscosity	$0.19 \pm 0.01^*$	$0.40 \pm 0.01^*$	$19.0 \pm 1.2^*$	$0.78 \pm 0.02^*$	$4.0 \pm 0.5^*$
Art. viscosity (with extra wall friction, see 4.3.3.2)	$0.19 \pm 0.01^*$	$0.38 \pm 0.01^*$	$15.5 \pm 0.5^*$	$0.82 \pm 0.02^*$	$5.0 \pm 0.7^*$

Table 4.5. Sloshing Experiment: Vertical Rod Bank (Test Case SD-D1R-2)

SD-D1R-2 (Dr = 17.6 cm)	Slosh at outer wall			Slosh at pool center	
	Arrival time at wall [s]	Time of maximum height [s]	Maximum height [cm]	Time of maximum height [s]	Maximum height [cm]
Experiment	0.20 ± 0.02	0.42 ± 0.02	15 ± 1	0.88 ± 0.04	15 ± 3
Artificial viscosity	$0.20 \pm 0.01^*$	$0.42 \pm 0.01^*$	$21.0 \pm 1.3^*$	$0.82 \pm 0.02^*$	$17.0 \pm 2.3^*$
Combined viscosity	$0.19 \pm 0.01^*$	$0.41 \pm 0.01^*$	$21.5 \pm 1.4^*$	$0.83 \pm 0.02^*$	$16.0 \pm 2.2^*$
Art. viscosity (with extra wall friction, see 4.3.3.2)	$0.20 \pm 0.01^*$	$0.41 \pm 0.01^*$	$18.5 \pm 0.6^*$	$0.84 \pm 0.02^*$	$15.5 \pm 2.1^*$

*The error range is defined using the percentage error values from the simulation of SA-D1-3 experimental case (see Table 4.2)

4.3.3 Discussion

4.3.3.1 General Flow Behavior

As can be seen from the visualization of the data, the shape of the flow in general is in good agreement with experimental observations. When released, the water column flows towards the rods. During its interactions with rods, the flow loses kinetic energy and the outward slosh is thus lower than in the test cases without rods. The main difference is that the shape of the outward slosh is wavier in the test case with rods. It can be seen that, in the numerical simulation, the areas between the rods and the container walls remain dry for a longer time than in the experiment. One possible explanation for this deviation is that the number of particles is not sufficient to catch all details of such a complicated flow pattern, and so the outward slosh consists of twelve high peaks, instead of the continuous wavy front.

At the same time, surface tension was not taken into account in the numerical model for this problem. The influence of surface tension on the bulk flow shape is very low. It can be shown that taking the diameter of the liquid column $d_c = 0.11$ m as a characteristic length and $v = 1$ m/s as a typical velocity, the Weber number will be approximately 1340 for the water under normal conditions. For the thin (0.001 m) and slowly moving (0.01 m/s) water layer at the wall of the container, the Weber number is of the order of 0.001, meaning that surface tension forces are dominant.

As a result of these limitations, the height of the calculated outward slosh is higher than the experimentally observed height, because former was measured using the heights of these twelve peaks. Nevertheless, there is no significant influence on the process of main interest – the formation of the central peak and the associated fuel compaction and possible recriticality.

For the test case with rods placed in the vicinity of the water column, the interaction of flow with the rod bank destroys the inward slosh. As a result, the central peak is very small. This result is of great importance for the accident scenario described in the introduction. From Table 4.4, it can be seen that calculated quantitative values are not in perfect agreement with the experimental data. The central peak height is well predicted, whereas the timing shows faster liquid movement. Implementing two models of viscosity, very similar quantitative results were obtained. However, the flow shape is more chaotic in the simulation with the combined viscosity model.

In the test case with rods placed near the container walls, the outward motion is similar to the first simulation case. The influence of the rods is smaller, because the liquid flows undisturbedly towards the center after hitting obstacles and reflecting from the wall, forming the central peak. However, the mass of the central peak and its height are smaller than in the case without the obstacles. Both viscous models predict similar quantitative results. They are in better agreement with the experimental observations, yet the calculated height of the outward slosh is nevertheless higher due to the different shape of the liquid front. The time of the maximum height is also a little bit lower than in the experiment.

4.3.3.2 Influence of Wall Friction

The issue of faster liquid movement presumably arises from the absence of a wall friction model. This is especially pronounced in the case of SD-D1R-1 with a rod bank located in the vicinity of the water column, as previously described. An additional complication is related to the container and rod surface conditions, which could not be properly reproduced in the series of experiments.

As boundaries in the numerical simulation are represented by "fixed" fluid particles, and as the first part of the term in the artificial viscosity is proportional to the difference of particle velocities, this relation can also be used for the friction simulation. Of course, this is not a physical model, but does allow the influence of the border friction on the liquid flowing through the rods to be checked. The α coefficient used for the viscosity simulation is relatively small. Thus, the idea is to also use the first part of the artificial viscous term for the wall friction, modeled by viscous interactions between fluid particles and wall particles with an increased value of α, denoted α_W. Two test cases – SD-D1R-1 and SD-D1R-2 – have been simulated using the following parameters: $\alpha = 0.01$, $\alpha_W = 0.1$ and $\beta = 0.1$.

The results of the simulations are shown in the bottom rows of Table 4.4 and Table 4.5. In both cases, the heights of the outward slosh are reduced. For test case SD-D1R-1, the outward slosh height agrees with the experimental data, while for test case SD-D1R-2, the height is still 2.5 centimeters higher than experimentally observed. However, the central peak heights are well reproduced in both cases.

4.4 Concluding Remarks

In this chapter, the SPH computing software has been applied to the numerical simulation of the three-dimensional sloshing liquid motion problem. A number of numerical models have been created to reflect different configurations of the experimental installations. These are as follows:

- Fully symmetrical configuration (the liquid column is symmetrically located at the container center);
- Asymmetrical configuration (the liquid column is located with an offset from the container center);

– Symmetrical configuration with obstacles (rod imitators are installed around the liquid column).

The sloshing motions of the liquid were well captured and reproduced by the numerical models. The details of the flow are better reproduced for the configurations without the obstacles (the first and second configurations in the list above), while for the complex geometry of the last configurations, some deviations in the shape of the slosh at the container wall were observed.

The quantitative parameters of the flows predicted by the numerical algorithm have been compared with the available results of the simulations performed with the SIMMER-III/IV reactor safety analysis code and with experimental data. These measured flow quantities, such as the heights of the wall sloshes and the central peak, and the timings of these events, are accurately predicted with high resolution simulations. At the same time, the present algorithm based on the SPH method is capable of resolving the high central peak in the fully symmetrical case, which was an issue for the SIMMER code.

A sensitivity study for the value of the central peak height in the symmetrical configuration has also been performed. The study showed the convergence of the central peak height value with an increase in the number of particles used for modeling.

In analyzing for a possible recriticality event, the height values of the central peak calculated for the different experimental configurations and different resolutions of the numerical model were compared. The highest peak, corresponding to the maximal volume of the fissile materials compacted in the center of the pool, is observed in calculations of the fully symmetrical configuration with the fine numerical resolution. Thus conclusion demonstrates the experimentally observed sensitivity of the liquid flow to the geometrical asymmetries of the vessel.

5 Parallelization of Numerical Code with CUDA Technology

5.1 Introduction

This chapter describes the parallelization of the serial numerical SPH code that was developed for the simulation of the centralized liquid sloshing experiment.

The parallelization of the SPH code is an important part of this study. The weakly-compressible approach (see Section 2.2) implemented in the present numerical model allows simulation of incompressible liquids, but at the cost of a relatively small value for the time step. The time step value depends on the average distance between particles, which reduces with increasing resolution of the system (number of fluid particles). At the same time, a large number of particles demands more computational power in order to calculate all the particle interactions in a single time step. These features of the weakly-compressible model in the SPH method result in a dramatically higher computational time to simulate a given time period of the sloshing experiment. For instance, simulation of the sloshing motion in symmetrical geometry with the number of particles $Np \sim 238 \times 10^3$, as described in Section 4.2, required more than two weeks on one core of an Intel Core 2 Duo 2.66GHz CPU.

One traditional way to perform parallel calculations is to incorporate changes into the source code of the numerical algorithm and to then run it on a multi-core machine. There are many approaches, such as OpenMP or MPI technologies, for performing parallelization of the numerical algorithm. Depending on the chosen technology of parallelization and on the algorithm type, the amount of programming effort required varies significantly. In the simplest of cases, it is limited to introducing special preprocessor directives at certain places in the source code. In other particular cases, deep reprocessing of the source code is needed.

An alternative way to perform parallel numerical calculations is *General-purpose computing on graphics processing units* (GPGPU). GPGPU is the technique of using graphics processing units (GPUs) for parallel numerical calculations. There are several modern technologies which allow the utilization of GPUs for parallel computation. They all have various levels of abstraction, allowing the programmer to either focus only on high level programming or to go deeply into the hardware level. The use of the GPU for parallel computations is very elegant; it allows high computational power from a regular desktop computer for a considerably lower price. Recently, several studies (see for instance

[35], [98], [99]) have demonstrated the high potential of the GPGPU approach for the computational fluid dynamics (CFD) problems.

In this chapter, the results of a comparison of the CPU and GPU runs of the reference cases are presented. The chapter concludes with a discussion on the speedups achieved in the sloshing simulations.

In Appendix B, a brief overview of the Compute Unified Device Architecture (CUDA) technology, used in the parallelization of the present SPH algorithm, is given. Following that, a description of the parallelization strategy is given, and the specifics of the current hardware are discussed. The final version of the parallel algorithm is then presented.

5.2 Results of the GPU Calculations

The parallelized SPH algorithm, described in Appendix B, has been used for the simulation of the following experimental cases:

- SA-D1-3 (symmetrical sloshing without rods);
- SD-D1R-2 (symmetrical sloshing with a rod bank imitator in the flow).

Firstly, these tests have been performed using the same parameters as for the CPU simulations. This is important, as it allows the results of the CPU and GPU simulations to be checked for differences, and makes it possible to evaluate the speedup achieved by parallelizing the code. Later, both experimental cases were simulated on the GPU with a higher number of particles (finer resolution).

No large differences in the flow shape have been observed between the CPU and GPU calculations for any of the simulated cases. The trajectories of some particles turned out to be slightly different, which could be a result of using single precision floating values. However, these small deviations in particle trajectories do not affect the general flow pattern.

5.2.1 Experimental case without rods

A plot comparing the CPU and GPU time for test case SA-D1-3 is presented in Fig. 5.1. The quantitative parameters of the numerical experiment are listed in Table 5.1. Direct comparison of the CPU time with the GPU time gives a speedup factor of 18.7. However, this value is not realistic for a number of reasons. The number of time steps performed to simulate one second of the real experiment is higher for the CPU simulation than for the GPU simulation. In the latest GPU calculations, the less conservative time stepping conditions are used. Another issue is that the calculation results for the CPU case were obtained with a not fully optimized final version of the code. Repeating the whole simulation with the final version of the serial code seemed to be unreasonable, as the estimated computational time is approximately two weeks. Thus, a coarse comparison has been made by measuring the CPU and GPU times on the first time steps of the test runs. The effective speedup was then recalculated using corrective factors (see also note to Table 5.1). The effective speedup in this test case was thus estimated to be 8.8.

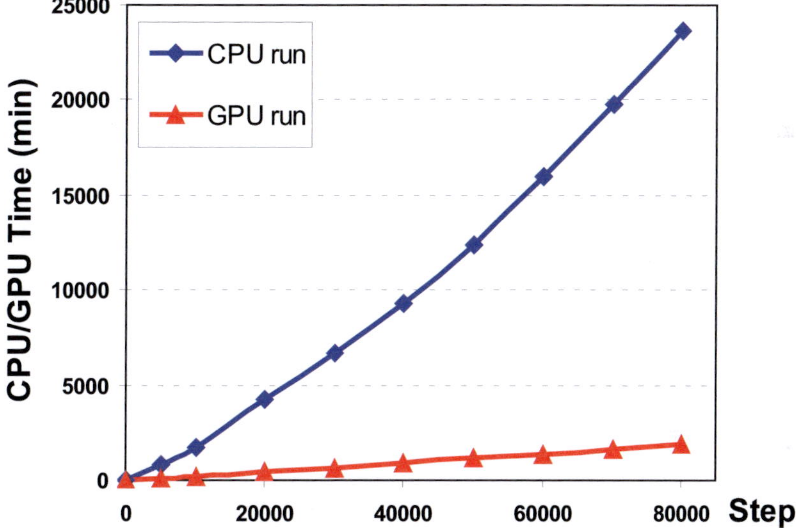

Fig. 5.1. CPU and GPU time in simulation of sloshing experiment. Experimental case SA-D1-3 ($Np \sim 238 \times 10^3$)

Table 5.1. Timings of the sloshing experiment simulation performed on CPU and GPU. Experimental case SA-D1-3 (Np $\sim 238 \times 10^3$)

SA-D1-3	CPU	GPU	Speedup	Effective Speedup
Simulation Time, min	26979	1430		
Real Time, ms	1000	1000	**18.9**	**8.8**
Number of Time Steps	88100	61860		

Note: The effective speedup was calculated dividing the "raw" speedup by 1.5 (since on the first time steps, the final CPU algorithm was approximately 1.5 times faster than the former one), and then multiplying by 61860/88100 to evaluate the time for the same number of time steps.

The most computationally expensive part of the parallel code is `GPU_SolveMomentAndContEq` kernel. Direct comparison of its execution time with the execution time for the similar part of the serial code gives a speedup of approximately 4–5. The difference with the overall speedup has a twofold nature. On one hand, the other parts of the numerical code work significantly faster in the parallel version. The averaged speedup of these parts is an order of magnitude greater (20–30, or as much as hundreds of time greater when the *memory copy* procedure is used) compared with the serial version.

On the other hand, the serial version slows down dramatically with increases in the number of interacting pairs of particles, and this leads to 2.5 times higher computational costs for one time step at the end of the simulation period than at the beginning. This calculational slowdown is clearly seen in Fig. 5.1: the slope of the curve related to the CPU run is increases highly first of all after approximately 5000 time steps from the beginning, and again after ~45000 time steps. This is more clearly seen in Fig. 5.2, where the computational costs for one time step are plotted for the serial CPU and parallel GPU versions of the algorithm.

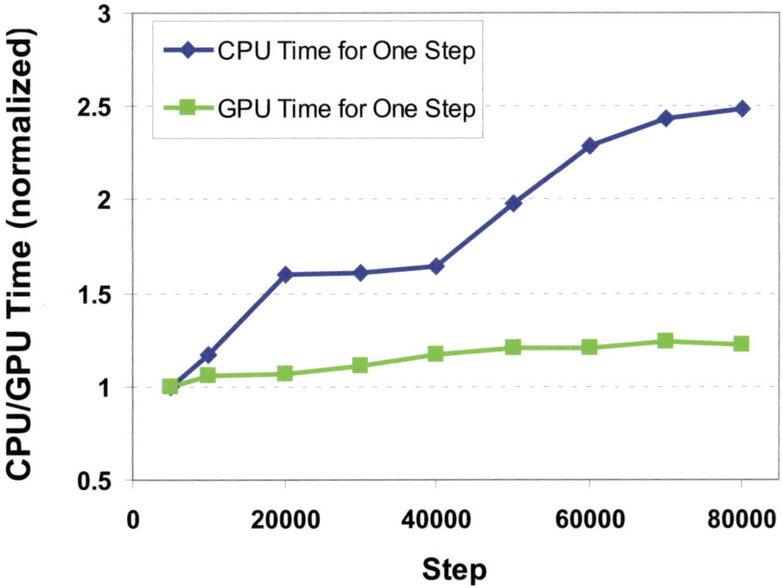

Fig. 5.2. Change in CPU and GPU time to perform a single time step. Experimental case

SA-D1-3 (Np $\sim 238 \times 10^3$)

For the parallel CUDA algorithm, the effect of the number of interacting particles is much less pronounced compared to the serial CPU algorithm, on account of the different architecture of GPUs. Without going deeply in the specifics of GPU internals, it needs to be taken into account that GPU devices of compute capability 1.3 and lower have no memory caching. Therefore, most of the time the GPU spends reading and writing data to and from memory, and only a fraction of the GPU time is spent performing arithmetical operations on data. A specific model of the GPU memory read/write procedure leads to a slightly increased computational time per time step at the end of the simulation, compared with the beginning. The final increase in computational costs is roughly 25%, according to the plot in Fig. 5.2.

5.2.2 Experimental case with rod bank

A second GPU simulation was performed for the experimental case SD-D1R-2. The geometry of the case includes a symmetrical bank of twelve rods installed in the vessel. The time plot of this simulation is shown in Fig. 5.3. For this case, the computational time of the serial algorithm was significantly higher than the parallel one. The quantitative performance parameters of the algorithms are given in Table 5.2. A direct comparison of the computational times gives a speedup factor of 12.3. This value has been recalculated in line with the analysis of the previous numerical experiment (see also note below Table 5.2), and the effective speedup value is estimated as 10.7.

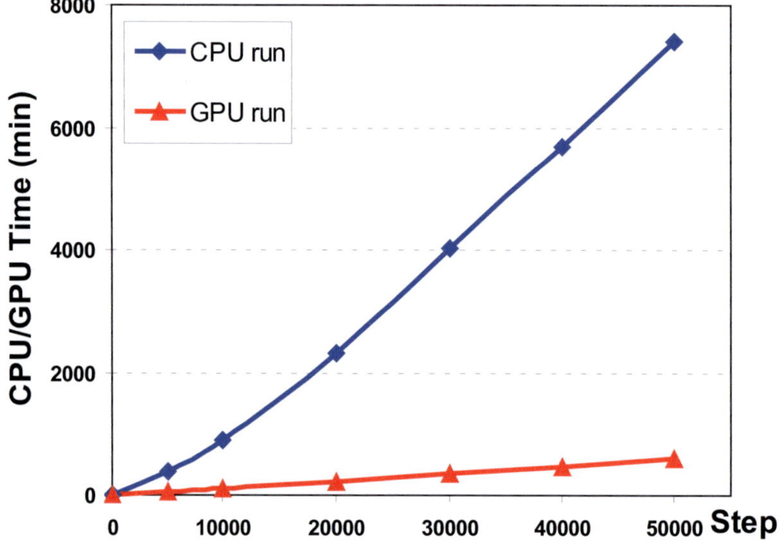

Fig. 5.3. CPU and GPU time in simulation of sloshing experiment. Experimental case
SA-D1R-2 $(Np \sim 128 \times 10^3)$

Table 5.2. Timings of the sloshing experiment simulation performed on CPU and GPU. Experimental case SA-D1R-2 (Np ~ 128×10^3)

SA-D1R-2	CPU	GPU	Speedup	Effective Speedup
Simulation Time, min	7721	628		
Real Time, ms	1000	1000	12.3	10.7
Number of Time Steps	~52000	~52000		

Note: The effective speedup is calculated dividing the "raw" speedup by 1.15 (as at the first step the modified CPU algorithm is approximately 1.15 times faster than the former one).

In this numerical experiment, a significant slowdown from the start to end of the serial algorithm also takes place, but one particular difference was observed. In contrast to the previous case, the slope of the curve for the CPU time of the serial algorithm changes once, starting from approximately 5000 and up to 30000 time steps. The explanation for this behavior is as follows. At this time point, the collapsing water column reaches the rod bank, resulting in an increasing number of interacting pairs of particles. Later, the particles representing the side wall of the container are also involved in interactions, resulting in an increase of the computational cost per time step. This process continues up to the moment when the slosh at the outer wall begins to collapse. At this moment, the number of interacting pairs becomes more or less stable, with only small deviations around the new value.

For the parallel algorithm, the computational cost per time step does not change dramatically, similar to the sloshing simulation without rods. At the end of simulation, the computational cost per time step is 15% higher than at the beginning. At one point, near 20000 time steps, the parallel algorithm is even slightly faster than at the beginning of the simulation. The reasons also lie in the specifics of the GPU architecture.

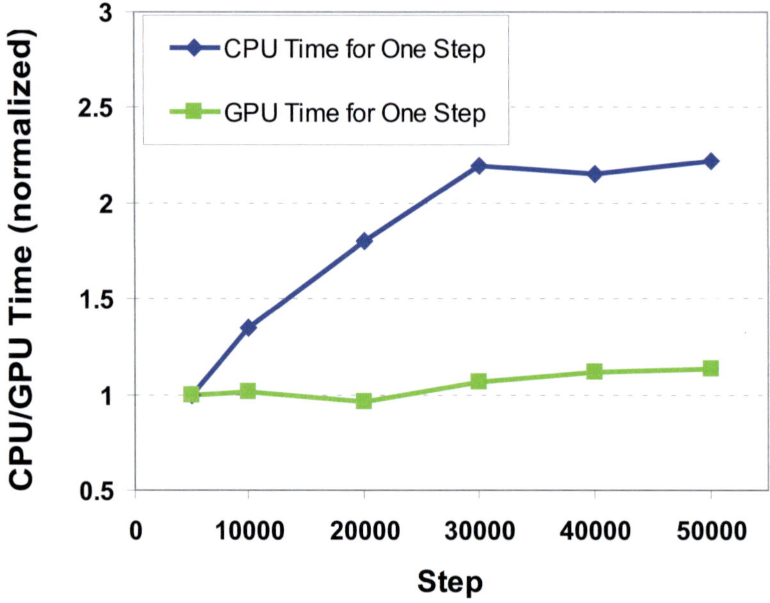

Fig. 5.4. Change in CPU and GPU time to perform one time step. Experimental case
SA-D1R-2 (Np ~ 128×10^3)

5.2.3 Simulations on GPUs with a compute capability 2.x

The speedups achieved with the parallel SPH code on the Tesla C1060 GPU are certainly very good. They are even higher than reported in [76], where SPH code was also parallelized on the CUDA technology. The most attractive point is that such performance was achieved on a standard general-purpose PC with a CUDA-enabled GPU. Unlike computational servers or supercomputers, this system does not need any special care. Needless to say, a standard PC with a GPU is much cheaper than a computational server.

On the other hand, the performance of the parallel code is far from the theoretically predicted peak performance for the GPU model used in simulations. There are several important aspects that can significantly slow down the performance of a CUDA-based algorithm. Such issues have been partially taken into account during development of the parallel version of the code. For instance, in the case of the algorithm for passing through neighboring cells, three versions of the algorithm logic were suggested. The specific of the shared memory and the amount of it installed on the GPU are the critical factors which define the version of the algorithm finally implemented in the code.

The factors that have a great impact on the performance of the CUDA algorithms are considered in guides to the CUDA technology [23, 24]. Following the recommendations in these guides, a couple of factors that significantly limit the performance of the developed code have been identified. They are:

- *Branching of the threads*: flow control instructions (if, switch, for, do, etc.) can significantly affect instruction throughput: when different threads in a warp (a group of 32 threads) are to perform different instructions, the algorithm is automatically serialized to perform all execution paths one by one;

- *Misaligned data pattern*: when the data is misaligned, reading is performed in a serial way, thread by thread. Since GPUs of compute capability 1.x have no cache for the global memory, the serialization of memory access significantly reduces performance.

The branching of the algorithm is usually difficult to avoid for the particular numerical model. A more promising way to increase the performance of the code is to optimize memory access. At first glance, the SPH method looks like a perfect candidate for parallelization, since the interactions of the particular particle with the neighboring particles can be calculated independently from all the other particles in the current time step. On the other hand, the particles move along with the evolution of the simulated system, and thus each particle's set of neighbors is changed at every time step. This results in a misaligned (random-like) memory access by the threads and thus significant delays in memory operations. It is neither clear how to reorder the particle data according to the current positions, nor if this would in fact lead to a higher performance of the overall code.

The problem of misaligned memory access arises in many scientific applications of the CUDA technology. Luckily, the GPUs of compute capability 2.0 and higher have global memory caches, which have the potential to increase speed of the memory reads for misaligned data. This idea has been examined by a test run of the experimental case SA-D1-3 (see Section 5.2.1) on a Geforce GTX 580 Standard Edition GPU. Brief technical specifications of this GPU are given in Table 5.3, in comparison with the Tesla C1060.

Table 5.3. Main technical data of Nvidia Tesla C1060 and Nvidia GTX 580.

	Tesla C1060	**GTX 580**
Compute capability (hardware version)	1.3	2.0
Number of cores	240	512
Device memory, GB	4.0 (Not cached)	1.5 (Cached)
Processor clock, MHz	1300	1544

The Geforce GTX 580 is not a high-performance calculation-oriented GPU, but a standard high-performance video card routinely used for graphical computations in computer games. Nevertheless, it has valuable computational power and, like other CUDA-enabled GPUs, can be used for numerical simulations. It has roughly twice many computational cores as the Tesla C1060. The global memory size is only 1.5 GB, but memory access is cached.

Due to its higher number of computational cores, the GTX 580 theoretically provides more than twice as much performance for algorithms with a large number of arithmetic operations and fully coalesced memory access. Analysis of the present SPH algorithm with a profiling tool shows that the GPU cores idle most of the time, waiting for data and/or for other threads in the same warp to finish their branches of the algorithm. In this case, computational advantage in the use of the GTX 580 can be obtained only from the cached memory access. A time plot of this simulation in comparison with the simulation performed on the Tesla C1060 is shown in Fig. 5.5, and the comparative speedup data for both GPUs are listed in Table 5.4.

Table 5.4. Speedups achieved in sloshing simulations on Tesla C1060 and Geforce GTX 580.

	CPU/GPU Time (min)	Speedup (over CPU)	Effective Speedup (over CPU)
Serial CPU	26979	-	-
Tesla C1060	1430	~18.9x	**~8.8x**
Geforce GTX 580	534	~50.5x	**~23.7x**

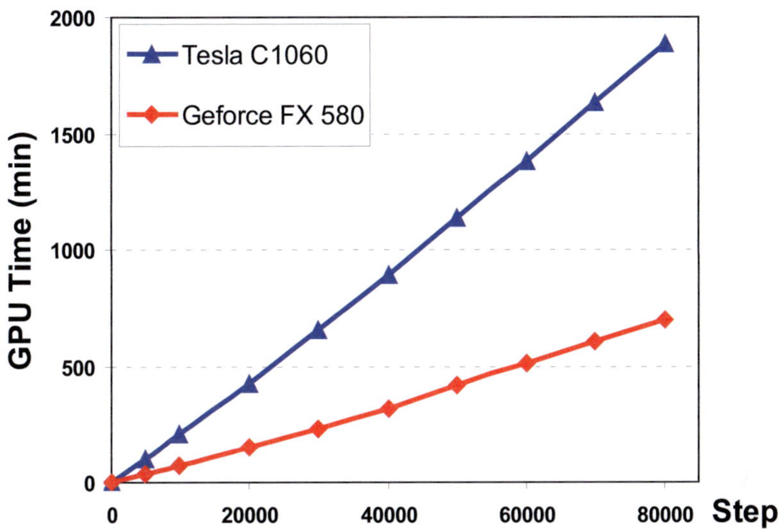

Fig. 5.5. GPU times in simulation of sloshing experiment on Tesla C1060 and Geforce FX 580 GPUs. Experimental case SA-D1-3 (Np $\sim 128 \times 10^{3}$)

The effective speedup is recalculated from the direct speedup value, following the same procedure as in the case of the results of the Tesla C1060 GPU (see note below Table 5.1 for the details). The resulting value of the effective speedup factor is 23.7 – almost three times higher than the speedup achieved in the calculation on the Tesla GPU. The change of GPU time per time step is plotted in Fig. 5.6. It can be seen to behave similarly to the Tesla GPU – the variations do not exceed 40%. At the beginning (between 5000 and 10000 steps), the performance of the GPU slightly decreases, and later increases again. Such a peak is not observed in the calculations on the Tesla GPU, and the reasons behind it are not clear.

To conclude the discussion on the speedups achieved, one additional note should be made. The maximum speedup using the parallel version of the code for the simulation of the sloshing motion without rods is estimated to be a factor of 24. The simulation of the experimental case with rods performed on the Tesla C1060 shows approximately 20% higher performance. These two facts allow it to be guessed that the simulation of the experimental case with rods in the flow on a Geforce GTX 580 GPU would show a speedup factor of about 28–30.

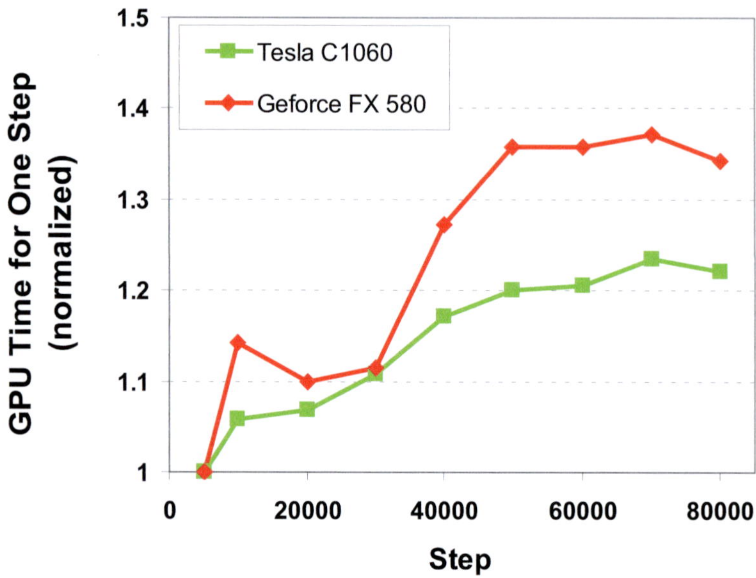

Fig. 5.6. GPU times spent in simulation of sloshing experiment on Tesla C1060 and Geforce FX 580 GPUs. Experimental case SA-D1R-2

5.3 Concluding Remarks

In this chapter (and partially in Appendix B), the development of a parallel module allowing the calculations to be run on graphical processing units (GPU) has been described. The parallelization of the SPH code was a key in establishing reasonable performance for the explicit time-integration scheme. The additional module has been developed using CUDA technology from Nvidia, which allows execution of the program on any CUDA-enabled GPU.

A number of speedup tests have been performed on several GPUs. The performance of the program on the standard general purpose desktop PC with the latest generation GPU is 25 times higher than on a single CPU computer without a GPU.

The GPU accelerated algorithm has been used for the high resolution runs of two configurations: the fully symmetrical case, and the symmetrical case with rod imitators installed around the liquid column. The results of the fine resolution calculations of the fully symmetrical case have been used in a sensitivity study for the central peak height value. Simulations of the latter case have made it possible to resolve the flow details around the rod imitators.

6 Conclusions

In the following, the main features and essential findings of this thesis will be summarized, along with proposals for future work.

In the current thesis, the Lagrangian meshless Smoothed Particle Hydrodynamics (SPH) method has been proposed as an appropriate numerical tool for CFD problems with complicated and fast changing geometries, of the kind that arise in nuclear reactor safety analyses. A numerical model based on the weakly compressible formulation of the SPH method has been developed.

The numerical model involves a number of known specialized techniques and approximations which improve the stability and accuracy of the original SPH approximation. One of the weak points of the original SPH method is the artificial viscosity model widely used in hydrodynamic simulations of free surface flows. To overcome this problem, a combined approximation of the viscous term in the momentum equation was introduced. The model allows approximating fluid viscosity using a dynamic viscosity, while at the same time maintaining the stability of the numerical solution.

The numerical model has been implemented as specialized computer software. The computing application has been developed following a modern programming paradigm – object-oriented programming – which makes the code flexible and easily upgradable. This is quite important for future studies on the application of meshless numerical methods to the simulations of nuclear-reactor-related problems. The application has a modern and complete graphical user interface to simplify interactions between the user and the software. The validity of the developed computing software has been carefully examined through a series of standard test problems: hydrostatic pressure distribution and the dam break problem.

The computing software has been applied to study a centralized sloshing liquid motion phenomenon, which can occur in accident sequences in fast liquid metal cooled nuclear reactors, in which the analysis of a hypothetical severe accident has to deal with a pool of molten material at the bottom of the reactor pressure vessel. The results of the numerical simulations have been verified against experimental data, and also against the results of computations obtained from the SIMMER-IV reactor safety code.

The meshless nature of the SPH method has turned out to be a remarkable asset in these numerical simulations. A number of advantages of the SPH method over traditional mesh-based numerical codes in particular free surface phenomena have been shown. Several different configurations of the sloshing liquid flows have been studied. The sloshing liquid motion has been well captured and reproduced by the numerical model. Important experimentally measured quantities, such as the central peak height and the timings of the sloshing events have been accurately predicted. The results of the numerical simulations confirm the experimentally observed sensitivity of the liquid flow to the geometrical asymmetries of the vessel. A dangerous condition for possible recriticality events with a high central peak has been observed only in strongly symmetrical configurations without any internal obstacles.

The most important issue in developing the computing software was to establish reasonable performance for the simulations. This is a challenge for the weakly compressible SPH method. To increase the speed of the calculations, an additional software module allowing the use of graphical processing units (GPU), instead of a central processor, has been developed. The module is written in CUDA C/C++ and allows the software to be run on any CUDA-enabled GPU. The software has been tested for the performance on several GPUs. The performance of the program on the PC desktop with a latest generation GPU is 20–25 times higher than on a standard general purpose single CPU computer.

One important conclusion of this work is that neither the SPH method nor other meshless methods should be neglected by developers of reactor physics simulation software, since the SPH modeling has a number of attractive features. The SPH method can be considered a possible alternative to the mesh-based methods traditionally used in nuclear engineering, and as a particular component among a variety of the numerical techniques used in complex reactor simulation software. Obviously, the primary application of the SPH method is to open boundary and free surface problems, as well as to problems with complex geometries, while the traditional mesh-based methods are clearly preferable in traditional, static geometry applications, on account of the SPH stability problems and higher computational demand.

For near future work, one obvious extension is to study the recriticality phenomenon in a configuration similar to the symmetrical sloshing experiment. To reach that goal, the developed software should be coupled with neutron kinetics code allowing criticality to be determined once the configuration during the evolution of a sloshing event is given. A number of calculations with different fuel compositions would demonstrate the likelihood of reaching a critical state in such configurations.

References

1. Adorni M., Bousbia-salah A., D'Auria F., Hamidouche T. Accident analysis in research reactors // Proc. of International Conference Nuclear Energy for New Europe 2007. Portoroz, Slovenia, September 10–13, 2007.

2. Adami S., Hu X.Y., Adams N.A. A new surface-tension formulation for multi-phase SPH using a reproducing divergence approximation // J. Comp. Phys. 2010, 229(13):5011–5021.

3. AIAA Guide for the Verification and Validation of Computational Fluid Dynamics Simulations (G-077-1998e), *AIAA Standards Series,* Published by AIAA, © 1998, 19 pages, Digital (www.aiaa.org).

4. Alberro J.G., Abbate A.D. Structural analysis of the reactor pool for the RRRP // Proc. of 18th Int. Conf. on Structural Mechanics in Reactor Technology (SMiRT-18). Beijing, China, August 7–12, 2005.

5. Andersen P.S., Astrup P., Eget L., Rathmann O. Numerical Experience with the Two-Fluid Model RISQUE // Proc. Topical Meeting, Thermal Reactor Safety, Sun Valley, Vol. 2 (1978).

6. Attia M.S., Meguid S.A., Liew K.M. Multiscale modelling of crush behaviour of closed-cell aluminium foam // Proc. of 2nd MIT Conference of Computational Fluid and Solid Mechanics, June 17–20, 2003, pp. 68–71.

7. Banerjee A., Hancock J.W. Higher order terms for a crack terminating at the interface between mismatched solids // Proc. of 2nd MIT Conference of Computational Fluid and Solid Mechanics, June 17–20, 2003, pp. 76–82.

8. Batchelor G.K. An introduction to fluid dynamics // Cambridge at the University Press, 1970. p. 615.

9. Becker M., Teschner M. Weakly compressible SPH for free surface flows // Proc. of Eurographics/ACM SIGGRAPH Symposium on Computer Animation. San Diego, California, USA, August 3–4, 2007, p. 209–217.

10. Bell C.R., Bleiweis P.B., Boudreau J.E., Parker F.R., Smith L.L. SIMMER-I: An Sn, Implicit, Multifield, Multicomponent, Eulerian, Recriticality Code for LMFBR Disrupted

Core Analysis // Los Alamos Scientific Laboratory report LA-NUREG-6467-MS, January 1977.

11. Belytschko T., Krongauz Y., Dolbow J., Gerlach C., 1998. On the completeness of meshfree particle methods // Int. J. Numer. Mech. Eng. 43: 785–819.

12. Bets V., Kriventsev V., An Object-Oriented Approach In Problems of Numerical Heat and Mass Transfer // Proceeding of 5th International Conference On Nuclear Engineering (ICONE-5), Nice, France, Log. No 2437, 1997.

13. Benz W., Asphaug E. Impact simulations with fracture. I. Method and tests // Icarus 107 (1), pp. 98–116.

14. Benz W., Hills J.G. Three-dimensional hydrodynamical simulations of stellar collisions. I-Equal-mass main-sequence stars // The Astrophysical Journal (1987) Vol. 323, pp. 614–628.

15. Brufau P., Garcia-Navarro. P. Two-dimensional dam break flow simulation // Int. J. of Num. Meth. in Fl., vol. 33, Issue 1, pp. 35–57.

16. Chen J.K., Beraun J.E., Jih C.J. An improvement for tensile instability in smoothed particle hydrodynamics // Comp. Mechanics. 1999. 23(4):279–287.

17. Cleary P.W. Modelling confined multi-material heat and mass flows using SPH // Appl. Math. Model. 1998. Vol. 22(12), p. 981–993.

18. Cleary P.W., Ha J., Prakash M., Nguyen T. SPH: A new way of modelling high pressure die casting // Proc. of 3rd Int. Conf. on CFD in the Minerals and Process Industries, CSIRO, Melbourne, Australia, Dec. 10–12, 2003.

19. Cleary P.W., Ha J., Prakash M., Nguyen T. 3D SPH flow predictions and validation for high pressure die casting of automotive components // Applied Mathematical Modelling. 2006. Vol. 30(11), p. 1406–1427.

20. Cherfils J.M., Blonce L., Pinon G., Rivoalen E. Towards the simulation of wave-body interactions with SPH // Proc. of 4th International SPHERIC Workshop. Nantes, France, May 27–29, 2009, pp. 173–179.

21. Colagrossi A., Landrini M. Numerical simulation of interfacial flows by smoothed particle hydrodynamics // J. of Comp. Phys., 191 (2003), p. 448–475.

22. Courant R., Friedrichs K., Lewy H. On the partial difference equations of mathematical physics // IBM Journal, March, 1967, pp. 215–234 // English translation of a paper originally published in Mathematische Annalen 100, 32–74 (1928).

23. CUDA C Best Practices Guide v.3.2 (20.08.2010) by Nvidia. http://developer.download.nvidia.com/compute/cuda/3_2/toolkit/docs/CUDA_C_Best_Practices_Guide.pdf

24. CUDA C Programming Guide v.3.2 (22.10.2010) by Nvidia. http://developer.download.nvidia.com/compute/cuda/3_2/toolkit/docs/CUDA_C_Programming_Guide.pdf

25. Cummins S.J., Rudman M. An SPH projection method // J. Comput. Phys. 1999. 152:584–607.

26. Duan R.-Q., Koshizuka S., Oka Y. Two-dimensional simulation of drop deformation and breakup at around the critical Weber number // Nucl. Eng. Des. (2003) pp. 37–48.

27. Duan R.-Q., Jiang S.-Y., Koshizuka S., Oka Y., Yamaguchi A. Direct simulation of flashing liquid jets using the MPS method // Int.J.Heat.Mass.Transfer 49(2006) pp. 402–405.

28. Dyka C.T., Randles P.W., Ingel R.P. Stress points for tension instability in SPH // Int. J. for Num. Meth. In Eng. Vol. 40, Issue 13, pp. 2325–2341, 1997.

29. Fatehi R., Manzari M. T. A consistent and fast weakly compressible smoothed particle hydrodynamics with a new wall boundary condition // International Journal for Numerical Methods in Fluids (2011), doi: 10.1002/fld.2586

30. Gingold R.A., Monaghan J.J. Smoothed particle hydrodynamics: theory and application to non-spherical stars // Mon. Not. R. Astr. Soc. 1977. 181:375–389.

31. Gingold R.A., Monaghan J.J. Binary fission in damped rotating polytropes // Mon. Not. R. astr. Soc (1978) vol. 184, pp. 481–499.

32. Gingold R.A., Monaghan J.J. Kernel estimates as a basis for general particle methods in hydrodynamics // J.Comp.Phys. 46 (1982) pp. 429–453.

33. Gonzales L.M., Sanchez J.M, Macia F., Souto-Iglesias A. Analysis of WCSPH laminar viscosity models // Proc. of the 4th International SPHERIC Workshop. Nantes, France, May 27–29, 2009, pp. 180–187.

34. Gopala V.R., Van Wachem B.G.M. Volume of fluid methods for immiscible-fluid and free-surface flows // Chem.Eng.J. Vol.141, 1–3 (2008) pp. 204–221

35. Harada T., Koshizuka S., Kawaguchi Y. Smoothed Particle Hydrodynamics on GPUs // Proc. of Computer Graphics International, 63–70 (2007).

36. Herault A., Vicari A., Del Negro C., Dalrymple R.A. Modeling Water Waves in the Surf Zone with GPUSPHysics // Proc. of 4th International SPHERIC Workshop. Nantes, France, May 27–29, 2009, pp. 77–84.

37. Hirt C. W., Nichols B. D. Volume of Fluid (VOF) Method for the dynamics of free boundaries // J. of Comp. Phys. 39 (1981) pp. 201–225.

38. Hu X.Y., Adams N.A. An incompressible multi-phase SPH method // J. Comput. Phys. 2007. 227:264–278.

39. Hu X.Y., Adams N.A. A constant-density approach for incompressible multi-phase SPH // J. Comput. Phys. 2009. 228(6):2082–2091.

40. Johnson G.R., Petersen E.H., Stryk R.A. Incorporation of an SPH option into the EPIC code for a wide range of high velocity impact computations // Int. J. of Impact Engineering. 1993. 14:385–394.

41. Jouniaux-Corriger Y., Izard J.P., Gilles P. Three dimensional fracture analysis of the reactor pressure vessel inlet nozzle under pressurized thermal shock // Proc. of 19th Int. Conf. on Structural Mechanics in Reactor Technology (SMiRT-19). Toronto, Canada, August 2007.

42. Kazimi M., Massoud M. A condensed review of nuclear reactor thermal-hydraulic computer codes for two-phase flow analysis // Energy Laboratory Report No.MIT-EL 79-018, February 1980.

43. Ketabdari M.J., Nobari M.R.H., Moradi Larmaei M., 2008. Simulation of waves group propagation and breaking in coastal zone using a Navier-Stokes solver with an improved VOF free surface treatment // Applied Ocean Research 30 (2008) 130–143.

44. Kolev N.I. IVA3: Computer code for modeling at transient three-dimensional three phase flow in complicated geometry // Karlsruhe Research Centre Report KfK 4950, Sept. 1991.

45. Koshizuka S., Ikeda H., Oka Y. Numerical analysis of fragmentation mechanisms in vapor explosions // Nucl. Eng. Des. 189 (1999) pp. 423–433.

46. Koshizuka S., Tamako H., Oka Y. A particle method for incompressible viscous flow with fluid fragmentation // Comput. Fluid Dynamics J. 1995. Vol.4(1), p. 29–46.

47. Koshizuka S., Oka Y. Moving particle semi-implicit method for fragmentation of incompressible fluid // Nuclear Science and Engineering. 1996. Vol.123, p. 421–434.

48. Leung A.Y.T., Su R.K.L. Two-level finite element study of axisymmetric cracks // International Journal of Fracture 89: 193–203, 1998.

49. Li W.H., Lam S.H. Principles of fluid mechanics // Addison-Wesley Publishing Company, 1964. p. 374.

50. Libersky L.D., Petscheck A.G., Carney T.C., Hipp J.R., Allahdadi F.A. High strain Lagrangian hydrodynamics – a three-dimensional SPH code for dynamic material response // J. of Comp. Phys. 1993. 109:67–75.

51. Liu M.B., Liu G.R., Zong Z., Lam K.Y. Numerical simulation of underwater explosion by SPH // In Atluri SN & Brust FW (Eds.): Advances in Computational Engineering & Science. 2000. p. 1475–1480.

52. Liu M.B., Liu G.R., Lam K.Y. Comparative study of the real and artificial detonation models in underwater explosions // Engineering Simulation. 2003. Vol. 25(2), p. 113–124.

53. Liu G.R., Liu M.B. Smoothed particle hydrodynamics. A meshfree particle method // World Scientific Publishing Co Pte Ltd, 2003, p. 449.

54. Liu M.B., Liu G.R., Lam K.Y. Constructing Smoothing Functions in Smoothed Particle Hydrodynamics with Applications // Journal of Computational and Applied Mathematics. 2003. Vol. 155, p. 263–284.

55. Liu G.R. Mesh free methods: moving beyond finite element method // CRC Press LLC, 2003, p. 693.

56. Lopez D., Marivela R., Garrote L. Smoothed particle hydrodynamics model applied to hydraulic structures: a hydraulic jump test case // J. Hydr. Res. (2009) 47:142–158.

57. Lucy L.B. A numerical approach to the testing of the fission hypothesis // Astron. J. 1977. Vol. 82, p. 1013–1024.

58. Martin J.C., Moyce W.J. An experimental study of the collapse of liquid columns on a rigid horizontal plane // Philosophical Transactions of the Royal Society of London.

Series A, Mathematical and Physical Sciences, Vol.244, No. 882 (Mar. 4, 1952), pp. 312–324.

59. Maschek W., Munz C.D., Meyer L. Investigations of sloshing fluid motions in pools related to recriticalities in liquid-metal fast breeder reactor core meltdown accidents // Nucl. Techn. 1992. Vol. 98(1), p. 27.

60. Maschek W., Roth A., Kirstahler M., Meyer L. Simulation experiments for centralized liquid sloshing motions // Karlsruhe Research Centre Report KfK 5090, Dez. 1992.

61. Monaghan J.J. Why particle methods work // J. Sci. Stat. Comput. 1982. Vol. 3, No. 4, pp. 422–433.

62. Monaghan J.J. Particle methods for hydrodynamics // Comput. Phys. Rep. 1985. Vol. 3, p. 71–124.

63. Monaghan J.J. On the problem of penetration in particle methods // J. of Comp. Phys. 1989. Vol. 82, p. 1–15.

64. Monaghan J.J. Smoothed particle hydrodynamics // Annu. Rev. Astron. Astrophys. 1992. 30:543–574.

65. Monaghan J.J. SPH without a tensile instability // J. of Comp. Phys. 159, pp. 290–311 (2000).

66. Monaghan J.J. Smoothed particle hydrodynamics // Rep. Prog. Phys. 2005. Vol. 68, p. 1703–1759.

67. Monaghan J.J., Cas R.A.F., Kos A.M., Hallworth M. Gravity currents descending a ramp in a stratified tank // J. of Fluid Mech., 379, pp. 39–69 (1999).

68. Monaghan J.J., Gingold, R. A. Shock simulation by the particle method SPH // Journal of Computational Physics. 1983. Vol. 52, p. 374–389.

69. Monaghan J.J., Huppert H.E., Grae Worster M. Solidification using smoothed particle hydrodynamics // J. Comp. Phys. 206 (2005): 684–705.

70. Monaghan J.J., Kajtar J.B. SPH boundary force // Proc. of 4th International SPHERIC Workshop. Nantes, France, May 27–29, 2009, pp. 217–220.

71. Monaghan J.J., Lattanzio J.C. A refined particle method for astrophysical problems // Astronomy and Astrophysics. 1985. Vol. 149(1), p. 135–143.

72. Monaghan J.J., Thompson M.C., Hourigan K. Simulation of free surface flows with SPH // Proc. of ASME Symposium on Computational Methods in Fluid Dynamics. Lake Tahoe, USA, June 19–23, 1994.

73. Morris J.P, Fox P.J., Zhu Y. Modeling low Reynolds number incompressible flows using SPH // Journal of Computational Physics. 1997. Vol. 136, p. 214–226.

74. Morris J.P. Simulating surface tension with smoothed particle hydrodynamics // Int. J. Num. Met. Fluids 2000, 33:333–353.

75. Nvidia CUDA Zone, http://www.nvidia.com/object/cuda_home_new.html.

76. Oger G., Jacquin E., Doring M., Guilcher P.M., Dolbeau R., Cabelguen P.L., Bertaux L., LeTouzé D. Hybrid CPU-GPU acceleration of the 3D parallel code SPH-Flow // Proc. of 5th Int. SPHERIC Workshop. Manchester, UK, June 23–25, 2010.

77. Okuda H. Nonphysical noises and instabilities in plasma simulation due to a spatial grid // Journal of Computational Physics 10 (3): 475 (1972).

78. Pigny S.L., 2010. Academic validation of multi-phase flow codes // Nuclear Engineering and Design 240, pp. 3819–3829.

79. Randles P.W., Libersky L.D. Smoothed particle hydrodynamics some recent improvements and applications // Comp. Meth. in App.Mech. and Eng. 1996. 138:375–408.

80. Repetto G., De Luze O., Birchley J., Drath T., Hollands T., Koch M.K., Bals C., Trambauer K., Austregesilo H. Preliminary analysis of the Phebus FPT3 experiment using severe accident codes (ATHLET-CD, ICARE/CATHARE, MELCOR) // Proc. of the 2nd European Review Meeting on Severe Accident Research (ERMSAR-2007). Karlsruhe, Germany, June 12–14, 2007.

81. Roach P.J. Computational fluid dynamics // Hermosa Publishers, Albuquerque, 1972, p. 434.

82. Rook R., Yildiz M., Dost S. Modeling transient heat transfer using SPH and implicit time integration // Num. Heat Transf., Part B, 51:1–23, 2007.

83. Roubtsova V., Kahawita R. The SPH technique applied to free surface flows // Computer & Fluids. 2006. Vol. 35, p. 1359–1371.

84. Sanders J., Kandrot E. CUDA by example: an introduction to general purpose GPU programming // Addison-Wesley, 2011. 290 p.

85. Schaffrath A., Fischer K.C., Hahm T., Wussow S. Validation of the CFD code FLUENT by post-test calculation of a density driven ROCOM experiment // Nucl. Eng. And Design. 2007. Vol. 237, p. 15–17.

86. Shadloo M.S., Yildiz M. Numerical modeling of Kelvin-Helmholtz instability using smoothed particle hydrodynamics // Int. J. Num. Meth. Eng. 87(10):988–1006, 2011.

87. Shao S., Lo E.Y.M. Incompressible SPH method for simulating Newtonian and non-Newtonian flows with a free surface // Advances in Water Resources 26 (2003) pp. 787–800.

88. Shepard D., 1968. A two-dimensional interpolation function for irregularly-space data // Proc. of the 23rd ACM National Conference, pp. 517–524.

89. Shirakawa N. R&D of the next generation safety analysis methods for fast reactors with new computational science and technology // Proc. of 14th JAEA-FZK/CEA/IRSN/ENEA SIMMER-III/IV Review Meeting. Karlsruhe, Germany, September 9–12, 2008.

90. SimSPH. User Guide // Karlsruhe Institute of Technology (KIT), Institute for Nuclear and Energy Technologies (IKET), 2011.

91. Slattery W.L., Benz W., Cameron A.G.W. Giant impacts on a primitive Uranus // Icarus, Vol. 99 (1992) 1, pp. 167–174.

92. Smith L.L., Sheheen N.N. SIMMER-II: A computer program for LMFBR disrupted core analysis // NUREG/CR-0453, LA- 7515-M, Rev. (June 1980).

93. Stoker J.J. Water waves // Interscience Publishers, New York, 1957.

94. Sussman M., Smereka P., Osher S. A Level Set Approach for Computing Solutions to Incompressible Two-Phase Flow // J. of Computational Physics, Vol. 114, 1 (1994) pp. 146–159.

95. Swegle J.W., Hicks D.L., Attaway S.W. Smoothed particle hydrodynamics stability analysis // J. of Comp. Phys. 116, pp. 123–134 (1995).

96. Takeda H., Miyama S. M., Sekiya M. Numerical simulation of viscous flow by Smoothed Particle Hydrodynamics // Progress of Theoretical Physics. 1994. Vol. 92, no. 5, p. 939–960.

97. Tanaka N., Ogawara T., Kaneda T., Maseguchi R. Numerical analysis of splashing fluid using hybrid method of mesh-based and particle-based modelings // Proc. of 13th

International Topical Meeting on Nuclear Reactor Thermal Hydraulics (NURETH-13). Kanazawa, Japan, September 27 – October 2, 2009.

98. Thibault J., Senocak I. Incompressible Navier-Stokes Solver Implementation on Single, Dual and Quad GPU desktop platforms with CUDA // 47th AIAA Aerospace Sciences Meeting, Orlando FL (2009), pap.no:AIAA-2009-758.

99. Tölke J., Krafczyk M. TeraFLOP computing on a desktop PC with GPUs for 3D CFD // Int. J. of Comp. Fl. Dyn.(2008), 22:7, pp. 443 — 456.

100. TOP500 list of the world's most powerful supercomputers (November, 2010), http://top500.org/lists/2010/11.

101. Vaughan G.L., Healy T.R., Bryan K.R., Sneyd A.D., Gorman R.M. Completeness, conservation and error in SPH for fluids // Int. J. Num. Meth. Fl. 2008; 56:37–62.

102. Veen D.J., Gourlay T.P. SPH study of high speed ship slamming // Proc. of 3rd ERCOFTAC SPHERIC Workshop on SPH Applications. Lausanne, Switzerland, June 4–6, 2008.

103. Verification Assessment, NPARC Alliance CFD Verification and Validation, http://www.grc.nasa.gov/WWW/wind/valid/tutorial/verassess.html.

104. Vignjevic R. Review of development of the Smooth Particle Hydrodynamics (SPH) method // Proc. of Dynamics and Control of Systems and Structures in Space (DCSSS), 6th conference, Riomaggiore, Italy, July, 2004.

105. Violeau D., Piccon S., Chabard J.-P. Two attempts of turbulence modelling in smoothed particle hydrodynamics // Advances in Fluid Modelling and Turbulence Measurements, World Scientific 2002, pp. 339–346.

106. Violeau D., Issa R. Numerical modelling of complex turbulent free surface flows with the SPH Lagrangian method: an overview // Int. J. Num. Meth. Fluids 53(2):277–304.

107. Vorobyev A., Kriventsev V. Smoothed Particle Hydrodynamics Method in Simulation of Liquid-in-Liquid Interaction // Proc. of 14th JAEA-FZK/CEA/IRSN/ENEA SIMMER-III/IV Review Meeting. Karlsruhe, Germany, September 9–12, 2008.

108. Vorobyev A., Kriventsev V. Particle method for liquid-in-liquid interaction simulation // Proc. of 13th International Topical Meeting on Nuclear Reactor Thermal Hydraulics (NURETH-13), Kanazawa, Japan, September 27 – October 2, 2009.

109. Vorobyev A., Kriventsev V., Maschek W. Analysis of Central Sloshing Experiment Using Smoothed Particle Hydrodynamics (SPH) Method (ICONE18-29805) // 18th International Conference on Nuclear Engineering ICONE18, Xi'an, China, May 17 – 21, 2010.

110. Vorobyev, A., Kriventsev V., Maschek W. Simulation of central sloshing experiments with smoothed particle hydrodynamics (SPH) method // Nuclear Engineering and Design 241 (2011) pp. 3086– 3096.

111. Waltar A.E., Reynolds A.B. Fast Breeder Reactors // Pergamon Press, 1980. p. 853.

112. Wood D. Collapse and fragmentation of isothermal gas clouds // Mon. Not. R. Astron. Soc. 1981. Vol. 194, p. 201–218.

113. Wu C., Huang G., Zheng Y. Theoretical Solution of Dam-Break Shock Wave // J. Hydraul. Eng. 125, 1210 (1999).

114. Xie H., Koshizuka S., Oka Y. Modelling of a single drop impact onto liquid film using particle method // Int. J. Numer. Meth. in Fluids. 2004. 45:1009–1023.

115. Xie H., Koshizuka S., Oka Y. Simulation of drop deposition process in annular mist flow using three-dimensional particle method // Nucl. Eng. Des. 235 (2005) pp. 1687–1697.

116. Xiong J., Arai J., Koshizuka S. Numerical analysis of droplet impingement with a thin water film on the wall using MPS method // Proc. of 13th International Topical Meeting on Nuclear Reactor Thermal Hydraulics (NURETH-13), Kanazawa, Japan, September 27 – October 2, 2009.

117. Yamano H., Hosono S., Sugaya M Analysis of sloshing experiments // Proc. of 14th JAEA-FZK/CEA/IRSN/ENEA SIMMER-III/IV Review Meeting, Karlsruhe, Germany, September 9–12, 2008.

Appendix A Computer Implementation of the Numerical Algorithm

A.1 Overview

The numerical model described in the previous sections has been implemented in as specialized computer software – SimSPH [90]. SimSPH is a computing application that allows the user to perform numerical simulations of fluid dynamics problems using the SPH method. The computational part of the application is unified for one, two, and three-dimensional problems. The number of dimensions is defined by the user in the initialization phase of the calculation.

The software was developed using Microsoft Visual C++ 2008 Express Edition, which is a freely distributed part of the Microsoft Visual C++ (MSVC) integrated development environment (IDE). Besides the attractive lack of expense associated with obtaining this development environment, Visual C++ allows the use of both standard (unmanaged) C++ and Managed C++ or C++/CLI, which are Microsoft extensions to the standard language, in the same application. The main part of the program is written in standard C++, and could be ported from the Windows operating system (OS) to other operating systems, such as Unix, Linux, etc. The computer program has an integrated graphical user interface (GUI) developed in C++/CLI. The GUI is not portable and works only on the Windows platform.

The software is delivered as an executable file (*SimSPH.exe*). When started, the application creates an output directory (./_OUTPUT) in the working directory. For every numerical simulation started by the user, the application creates a separate directory for the output information (./_OUTPUT/current_exp). If the directory with that name already exists, the newly created directory would be renamed using current date and time in the name.

The software can use any input files located in the input directory (./_INPUT). Usually, this option is only employed for numerical simulations with an intermediate stop, since the initial configuration of the numerical problem is done with the GUI. The input file has the same format as the output files with particle data, which are produced periodically during the program run.

A.2 Object-oriented programming model. Class hierarchy

Many of the numerical codes dealing with reactor physics were developed using a structured programming paradigm. Without going into details, it would be appropriate to say that structured programming is focused on procedures. A program written under the structured programming paradigm may contain one long list of instructions, or a group of instruction lists (called *subroutines* or *functions*).

The more straightforward and modern way to write computer programs is the object-oriented programming (OOP) model. Nowadays, this approach is widely used to develop many different types of computer software, including CFD software for scientific calculations [12]. This paradigm was employed in developing the software presented in this section.

OOP-based programs do not deal with procedures working with data in memory, but rather with objects, which include data and the procedures operating on this data. Complex types of the objects are commonly called *classes*. A crucial stage in developing an OOP-based program is to design the structure of the classes and the relations between them – the *class hierarchy*.

The class structure at the base of the software presented here begins with very basic data types. One of the simplest classes is the class of vector values. The base class of the vector values is the TVector class, which describes vector values in N-dimensional space. It also includes a number of methods for initializing vector objects and a number of service methods. For the practical implementation of three-dimensional vector values, the subclass TVector3D is derived from the TVector class. This class is extended with common methods operating on vector values, such as a method computing the length of the vector. The TVector3D class is widely used in the description of other classes. For instance, one of the main classes is the TParticle class, which describes mathematical particles as complex data-objects. Every instance of the TParticle class includes 8 instances of the TVector class:

- Coordinate (at the current time step);
- Coordinate (from previous time step);
- Velocity (at the current time step);
- Velocity (from previous time step);
- Velocity after applying XSPH procedure (at the current time step);
- Velocity after applying XSPH procedure (from previous time step);
- Acceleration (at the current time step);
- Acceleration (from previous time step).

Using the TParticle class as a base, more complex data-structures, such as a class of objects defining fluid volumes, have been developed. At the top of the class hierarchy is the TProblem class, which describes the numerical problem under, study including all parameters of the numerical model and all the particles involved in the simulation. Only one object of this class exists during the program's execution. In addition to the classes related to the numerical algorithm, classes for visualizing the current state of the simulated system were also developed. A full list of implemented classes, with a short description of each, is given in Table A.1.

Table A.1. Hierarchy of classes

Class Name	Source File	Short Description
TVector	*Vector.h*	A base class for all vector quantities in N-dimensional space
TVector3D (inherits from TVector)	*Vector.h*	A subclass derived from the TVector class. Defines three-dimensional vector quantities
TParticle	*Particle.h*	A base class for fluid particles
TBorderParticle	*Particle.h*	A base class for border particles
TParticleArea	*Particle.h*	A class for defining liquid (gas) volumes. Consists of a number of particle objects (members of TParticle class) and general properties and procedures
TBorderArea	*Particle.h*	A class for defining solid walls. Consists of a number of border particles (members of TBorderParticle class) and general properties and procedures

TProblem	*Particle.h*	A superclass for defining a numerical problem. Includes all problem-related parameters. May consist of a number of objects of TParticleArea class-type and TBorderArea class-type, one object of TAddMesh class-type, one object of TInit class-type. During run-time, only one instance of this class exists
TAddMesh	*AddMesh.h*	A class for defining the additional mesh used in the neighbor search algorithm. Includes a number of objects of TCell class-type and some data-managing procedures
TCell	*AddMesh.h*	A base class for the cells of the additional mesh
TInit	*Init.h*	A class for defining input and output data-objects for the objects of TProblem class-type. The objects of this class are used for storing the current state of the numerical experiment
TVisualTool	*VisualTool.h*	A base class for the objects used in visualizing the current state of the numerical model
TVisualToolVS (Inherits from TVisualTool)	*VisualToolVS.h*	A subclass derived from the TVisualTool class. It is an implementation of the parent class with the tools of Microsoft Visual Studio. Only one object of this class exists during run-time

The definitions of the different classes are placed in separate source files, in line with the concept of modularity, which allows one part of the program to be changed independently from others. This program modularity is very important to allow the software to be further developed and modified by different people, as is usually the case for specialized software for scientific numerical computations.

A.3 Graphical User Interface

It is very clear that the usability of the computer program depends highly on the functionality of the GUI. The GUI for the presented software has been developed taking into account the fact that some operations that are a native part of every numerical simulation are in fact rather annoying for the user. Most of specialized numerical codes in reactor physics are based on the following working concept: "preparation of input – simulation (waiting) – processing of output – analyzing the data." Such a sequence is the result of the limitations of the development tools formerly available. Nowadays, computer development tools and computer hardware allow the creation of very complex software systems. Thus, for numerical code, it is very desirable to automate such operations as the generation of input files, the processing of output data, and the visualization of numerical results.

The graphical user interface (GUI) has been developed as a part of the computational software (such an approach is also known as a *rich client application*). The integrated GUI allows the user to perform in an intuitive way the standard routine procedures common to all computational software. A weak point in the integration of the GUI into the application is that both parts are not entirely independent of each other, so the application cannot be ported easily to another platform. The complete separation of the GUI from the main computational part has not yet been carried out.

The main functionality of the GUI can be described by the following points:

- Creation of the computational model during the program run, without the use of input files. The user defines a type of the numerical problem and its main geometrical and physical parameters;

- The numerical experiment is controlled by a number of buttons, such as "RUN," "STOP," "PERFORM ONE TIME STEP," "RUN UP TO," which allow the user to examine the simulated numerical problem;

- The output parameters and the output data-file formats can be easily tuned, including during run-time;

- The current state of the numerical model is visualized during run-time, which allows the state of the model to be inspected and possible problems to be fixed during the initial stage of the computations. This is very important for long simulations;

– In addition to the previous note, the quantitative data about the current state of the whole system – as well as about every single particle – are available to the user at every time step.

In Fig. A.1, an overview of the main window of the running program is given. The GUI can be divided in several parts. The usual menu bar is located at the top of the main window. Below are positioned three toolbars. The first is the common toolbar used to create a new numerical problem, to save screenshots, to save the current experiment or particle data, and to refresh the visualization. Two other toolbars control whether the numerical experiment is performed on the CPU or GPU. The GPU-controlling toolbar contains the same control buttons as the CPU toolbar, plus several control buttons for transferring data to and from GPU memory.

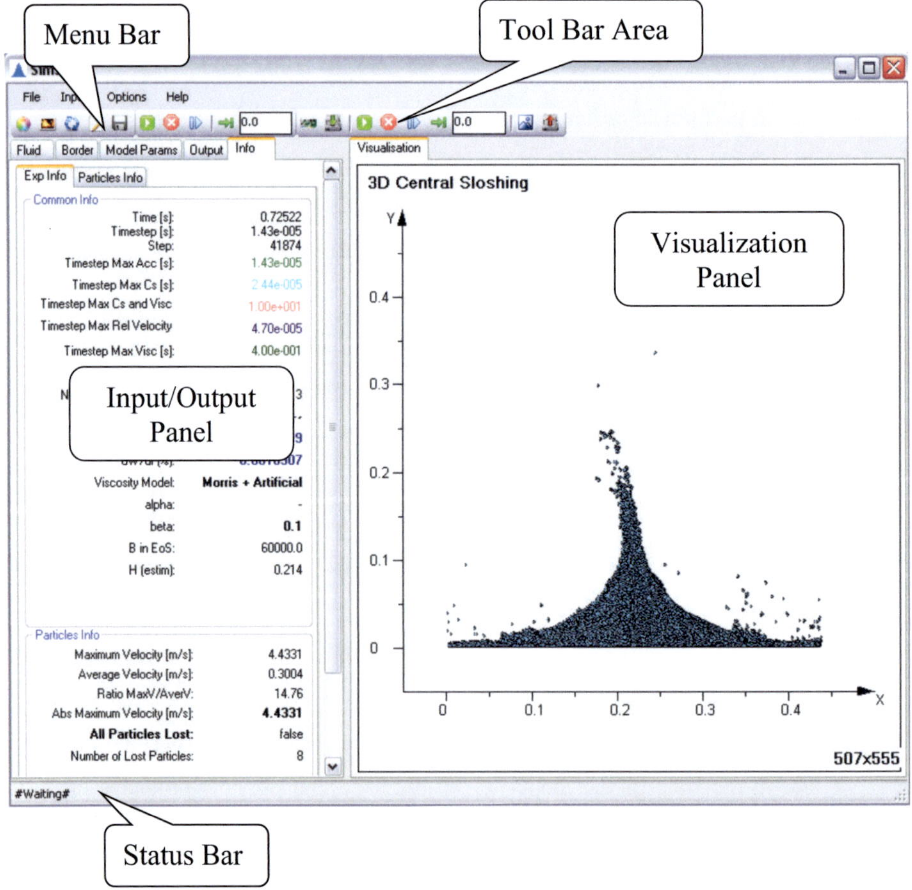

Fig. A.1. Graphical user interface. Main window

The main central part of the window is separated into two panels. The left panel is used to enter input data, to control the output produced by the program, and to check the current experimental parameters. It has a number of tabs for different tasks. These are as follows:

1. *Fluid Tab* – to configure fluid or gas volume(s) involved in the numerical simulation;

2. *Border Tab* – a tab for configuring the parameters of solid walls;

3. *Model Params Tab* – a tab holding the common model parameters, such as the model of viscosity employed, the parameters for density reinitialization, time integration parameters, etc.;

4. *Output Tab* – used to configure the output data, as well as the visualization of the current state of the computational domain;

5. *Info Tab* – this tab includes two parts: one shows the current experimental parameters, while the other is used to check the particle data by user choice.

The right-hand panel is used to display the visualization of the current state of the particle system. The visualization is two-dimensional. For three-dimensional problems, it shows a side-view projection of the particles.

The lowest part of the main window is occupied by a status bar, which indicates important information about the current state of the software.

The GUI also includes a couple of dialog windows and a console window. The console window is launched along with the main window and shows information about current operations being performed by the software. The standard Windows dialogs appear only when user chooses the input or output file for loading and saving data.

Appendix B Description of the Parallel Algorithm for GPU

B.1 CUDA Technology. Overview

The Compute Unified Device Architecture (CUDA) is a parallel computing architecture developed by Nvidia Corporation [75]. This technology allows developers to easily gain access to the computational power of graphical processing units based on Nvidia chipsets. The introduction of the CUDA technology was a response to the increasing interest of researchers into using the GPU, not as a device to perform graphical calculations, but as a powerful mathematical coprocessor for massively parallel scientific calculations.

The main idea of CUDA technology is to give developers a convenient and powerful tool for parallelizing numerical code and running it on GPUs. This goal has been achieved with the development of a new programming language, CUDA C/C++, an extension of the standard C/C++ language. After its initial release in 2006, the technology continues to develop rapidly: later releases of the CUDA language are almost fully compatible with a native C/C++ language, including support for the object-oriented programming model, while earlier releases provided only limited compatibility with standard C/C++.

The CUDA technology is not only a special programming language, but a whole system that includes:

- <u>Hardware:</u> GPUs with Nvidia chipsets for a wide range of computers, such as desktop computers (PCs), multicore and multiprocessor computational servers, and even supercomputers;[1]
- <u>Software:</u> CUDA Toolkits, including a C/C++ compiler (*nvcc*) with the support of the CUDA extension, developer tools, additional libraries, and third-party compilers for the other languages, such as Fortran, allow parallelization of older code not written in C/C++ language.

Compared to the former releases of GPGPU, the modern CUDA is a scalable programming model. This means that a numerical code written in CUDA C/C++ will work on other CUDA-

---

[1] One of the most powerful supercomputing systems in the world – the Nebulae system in <u>National Supercomputing Centre in Shenzhen (NSCS)</u> – is equipped with 64960 Tesla C2050 GPUs, and is third in the ranking of the fastest supercomputers in the world (as of November 2010) [100]

enabled GPUs with different numbers of cores. This feature makes programming for GPUs similar to programming for an ordinary machine with a single or multi-core CPU.

Another important feature of the CUDA programming model is the heterogeneous serial-parallel programming. The heterogeneous approach allows code to be split into a serial part, which is performed on a CPU, and a parallel part, which is performed on a GPU. Both can work individually or, when needed, simultaneously. This feature of the CUDA model gives great benefit to code originally written in the C/C++ language according to the modular programming model, as is the case for the serial numerical code described in Appendix A.

These features of CUDA technology, which perfectly match the architecture of the SimSPH code, have helped to make the choice of this technology for parallelization.

B.2 Description of the Parallel SPH Algorithm

The CUDA parallelization strategy for numerical code depends on a number of code parameters:

- The programming language or languages used in the code development – code written in the C/C++ language can be naturally coupled with parallel CUDA code and compiled using the *nvcc* compiler. Other languages require the use of third-party compilers or wrappers;
- The modularity of the code; this decides on the possibility of dividing the code into logically independent parts. A high level of modularity leads to small changes in the code, since the part to be parallelized can be easily set apart, while the main part of the code remains unchanged;
- The kind of numerical algorithm implemented in the numerical code.

For the present SPH code, the C/C++ programming language was employed. This made possible the use of the *nvcc* compiler and allowed code written in standard C/C++ to be merged with a newly developed parallel module.

In addition, the present code has a high level of modularity. All functions which are parts of the SPH solver are located in a separate file, *SPH_solver.cpp*. Therefore only the functions assembled in this file need to be rewritten in the CUDA C/C++ language.

Finally, the SPH algorithm described in Section 2 is almost fully serial. The interactions of every *i-j* pair of particles can be calculated independently from the interactions of other pairs.

These features of the developed serial SPH code made the parallelization relatively simple. Practically, it was performed in several steps. First, an additional CUDA source file, *SPH_solver_GPU.cu*, was added to the Visual Studio project. All functions were transferred to this new source file, and then transformed to so-called *kernels*. To avoid possible confusion, it is important to distinguish the smoothing kernels used in the SPH approximations, from these kernels, which are the CUDA functions executed in parallel on the GPU. The CUDA kernels have the same syntax as standard C/C++ functions, but with several additional parameters in function calls named *execution configuration* – parameters defining the number of threads and the number of blocks of threads to be executed in parallel. The functions that were transformed into CUDA kernels are listed in Table B.1.

In addition to the CUDA kernels, a number of auxiliary functions were added to the project. Since the CPU and GPU each have their own memory (called *host memory* and *device memory*, respectively), the particle data should be transferred to the device memory before any operations can be performed on these data. Moreover, periodical copies of the data are required to create output files and to support visual output to the GUI. The functions performing data copying between the host and device memories are also listed in Table B.1.

As mentioned before, CUDA technology continues to develop rapidly. The latest generation of CUDA enabled GPUs provides support for a number of new features. On other hand, the older generation of GPUs has several limits in respect of some important operations. The parallel code described in this section was developed to run on a Tesla C1060 GPU, which has a so-called compute capability 1.3 (this number is actually a version number of the CUDA-enabled hardware). The main limitations of this generation of GPUs are that:

- Double precision floating-point operations are much slower than single precision operations;
- The classes of the C++ language are not supported;
- Only a limited number of atomic operations is supported.

These limitations have defined the strategy of parallelization and the architecture of the parallel code. As floating-point operations with double precision are too slow on the available GPU, it was decided to perform GPU calculations with single precision accuracy. The corresponding transformation of variable types is performed as the data is copied from the host to the device and back.

Table B.1. List of CUDA kernels and auxiliary functions developed during parallelization of
the code

		Routine name	Description
GPU Functions (Kernels)	**Transformed from CPU functions**	GPU_RefreshMesh2D	Assigns particle indices to the cells of the additional mesh (for 2-dimensional problems)
		GPU_RefreshMesh3D	Assigns particle indices to the cells of the additional mesh (for 3-dimensional problems)
		GPU_XSPH	Performs XSPH corrections of particle velocities
		GPU_SolveMomentAndContEq	The SPH solver; calculates new values of particle accelerations and rates of density change
		GPU_ReInitDensity	Updates densities following the reinitialization procedure
	New	GPU_GetMaxNumPartInCell	Calculates the number of particles in the cells of the additional mesh
		GPU_FindMaxAcceleration	Searches for a particle with maximum acceleration
		GPU_PredictorUpdateParams	Updates particle parameters on the predictor half-step
		GPU_CorrectorUpdateParams1	Updates particle parameters on the corrector half-step (first)
		GPU_CorrectorUpdateParams2	Updates particle parameters on the corrector half-step (second)
CPU Functions	**New**	CopySPHDataToGPU	Allocates GPU memory and copies problem data to the GPU device
		CopyRunDataToHost	Copies current parameters of the run back to the host memory (current time, number of time steps, etc.)
		CopyParticleDataToHost	Copies memory parameters of particles back to the host in order to create the output
		CopySPHDataToHost	Copies all data back to the host memory and frees memory on the GPU device

The limited support for C++ classes was a severe issue for the parallelization, because the serial code was written according to the object-oriented approach, which widely uses classes (see Appendix A for details). On the other side, the performance of the CUDA algorithm, which utilizes "structures of arrays," is faster than the performance of the CUDA algorithm based on the "arrays of structures" approach, due to misaligned memory access [23]. As a result, the data structures in the parallel algorithm have been redesigned following the "structures of arrays" approach. Practically, this means using arrays of coordinates and other parameters rather than arrays of objects (particles) each of which is assigned coordinates and other parameters.

Finally, the limited number of atomic operations with floating-point value resulted in the development of the kernels for finding maximal acceleration over all particles (GPU_FindMaxAcceleration) and for finding a cell of the additional mesh with the maximal number of particles (GPU_GetMaxNumPartInCell). In the serial code, these operations are quite simple. The highest acceleration is found in the main cycle through all the particles in the SPH solver by comparing values of acceleration particle by particle. A similar procedure is used to get the maximal number of particles in cells of the additional mesh, checking cells one by one. However, these simple operations become more complicated when performed on a parallel machine. On a parallel machine, many threads together may compete to access the same data, thus producing the wrong results or even failure of the calculation in the worst case [84]. Atomic operations have been especially developed to avoid this problem, but unfortunately GPUs with compute capability 1.3 do not support atomic operations on floating-point numbers. As a result, the search for maximal values is performed in special kernels where the data to be analyzed is distributed between parallel threads, so that two threads never access the same memory address.

Finally, significant changes have been made to the SPH solver. To define the most effective way of passing through the particles, three versions of the XSPH procedure have been developed and compared to the test cases. The fastest way was then implemented in two other kernels: GPU_SolveMomentAndContEq and GPU_ReInitDensity. The logic of the algorithm for passing through the cells of the additional mesh for the given particle is schematically presented in Fig. B.1.

The algorithm is as follows: initially, for the given particle i, a block of threads is executed. The first thread of the block calculates the interaction between the given particle and the first particle in one of the neighboring cells. The first thread then moves to another neighboring cell and calculates the interaction between the given particle and the first particle in the current neighboring cell. This process is repeated until the first thread has passed through all the neighboring cells, including the cell where the given particle is located (interactions calculated by the first thread are shown by dashed lines in Fig. B.1). At the same time, the second thread calculates the interactions between the given particle and the second particles in the neighboring cells (partially illustrated by the dashed-double-dotted arrows in Fig. B.1). When all threads in a thread block finish cycling through the neighboring cells, all the interactions of a given particle with the neighboring particles are calculated and stored in shared memory (that is, common memory for all the threads of the same block). Finally, the first thread sums the data for the given particle while other threads of the same thread block are idle.

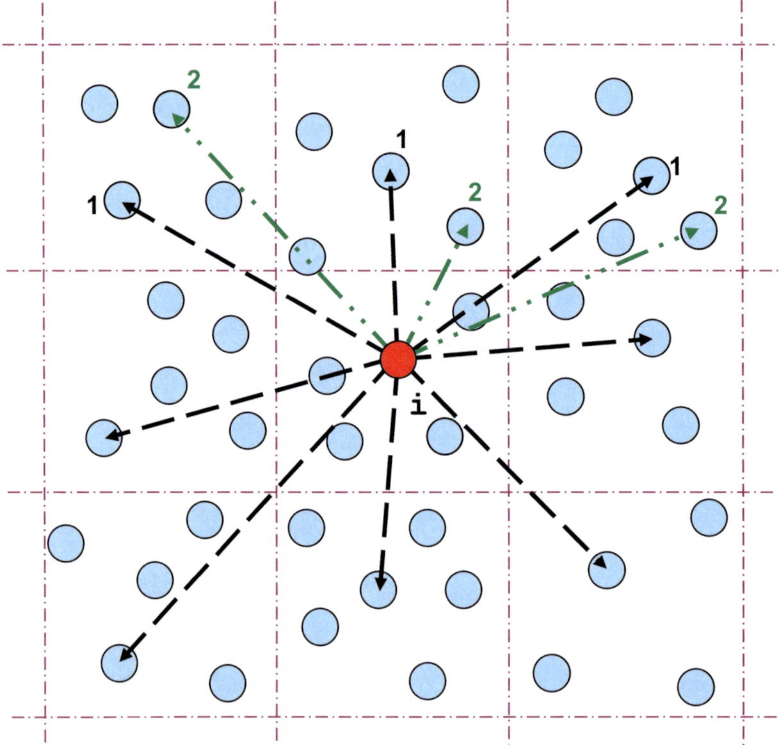

Fig. B.1. The algorithm for searching for interactions between the given particle and neighboring particles. Dashed arrows are interactions calculated by the first thread; dashed-double-dotted arrows are interactions calculated by the second thread

The size of the thread block should be equal or higher than the maximum number of particles in the neighboring cells. As threads on the GPU are executed in so-called *half-warps* – a group of 16 threads – the number of threads in the thread block is rounded up to the nearest multiple of 16. A scheme illustrating the algorithm for passing through the neighboring cells of the additional mesh by the threads of the ith thread block is schematically illustrated in Fig. B.2.

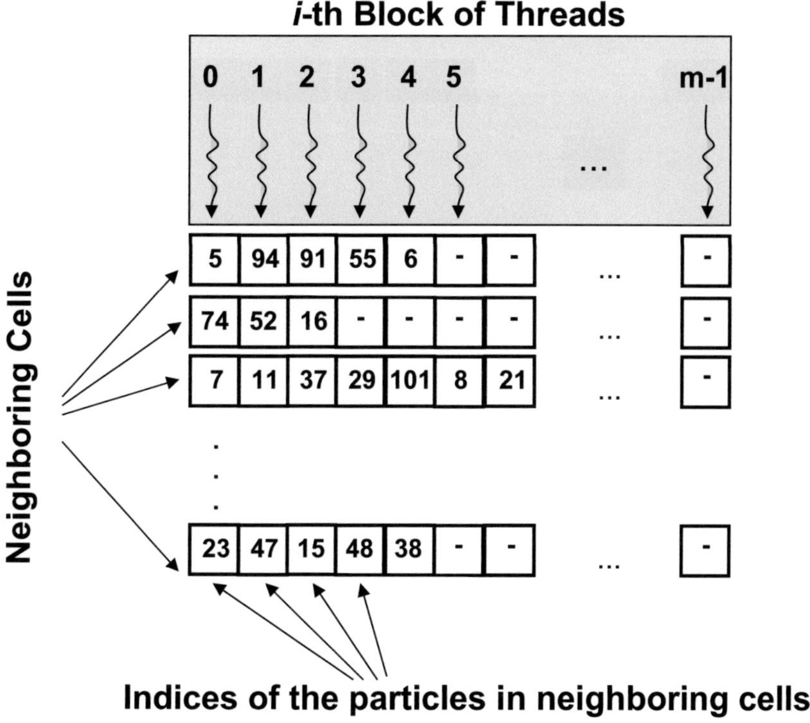

Fig. B.2. Algorithm for passing through the cells of the additional mesh. The numbers in squares are indices of particles located in the corresponding cell

To calculate the interactions in a system of N particles, the corresponding number of thread blocks must be run. For GPU devices with compute capability 1.3, the size of a grid of thread blocks is defined by a one or two-dimensional value. The size should not exceed 65535 thread blocks in the x-dimension or the y-dimension [24]. It is worth mentioning that the number of particles in 3D simulations is usually higher than the limiting value. For this reason, the execution configuration for the solver's kernels are defined by a two-dimensional grid of blocks and a one dimensional block of threads.

The parallel code is executed on both CPU and GPU. The CPU part controls the parameters of the run, starts the GPU kernels, performs periodical outputs, and produces the visualization of the current state of the particle system. The GPU part is used only for the execution of the kernels. The overall parallel algorithm is shown in Fig. B.3.

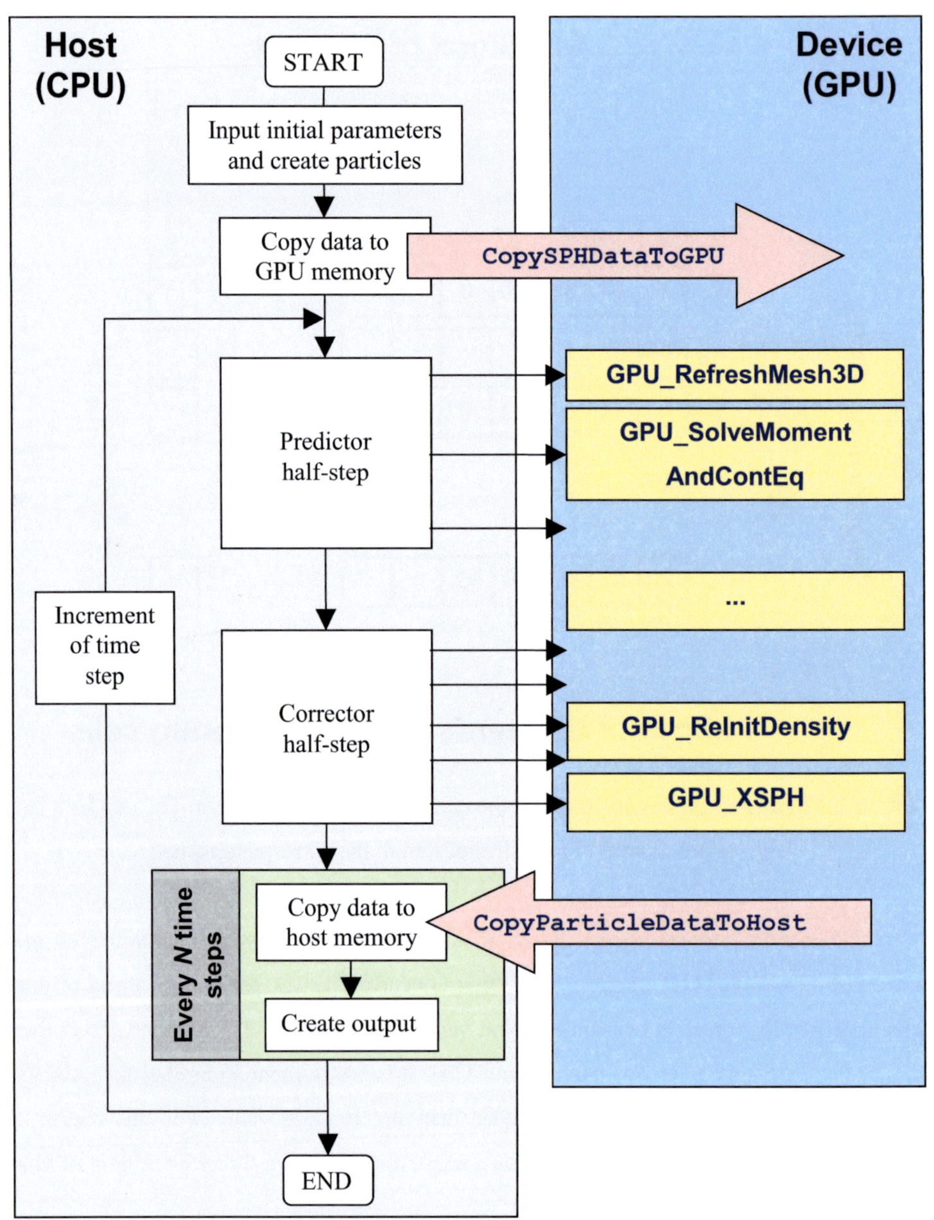

Fig. B.3. SPH algorithm parallelized with CUDA technology